THE ENVIRONMENTAL HEALTH CRITERIA

Environmental Health Criteria 230

NITROBENZENE

First draft prepared by L. Davies, Office of Chemical Safety, Therapeutic Goods Administration, Australian Department of Health and Ageing, Canberra, Australia.

Published under the joint sponsorship of the United Nations Environment Programme, the International Labour Organization and the World Health Organization, and produced within the framework of the Inter-Organization Programme for the Sound Management of Chemicals.

World Health Organization
Geneva, 2003

The **International Programme on Chemical Safety (IPCS)**, established in 1980, is a joint venture of the United Nations Environment Programme (UNEP), the International Labour Organization (ILO) and the World Health Organization (WHO). The overall objectives of the IPCS are to establish the scientific basis for assessment of the risk to human health and the environment from exposure to chemicals, through international peer review processes, as a prerequisite for the promotion of chemical safety, and to provide technical assistance in strengthening national capacities for the sound management of chemicals.

The **Inter-Organization Programme for the Sound Management of Chemicals (IOMC)** was established in 1995 by UNEP, ILO, the Food and Agriculture Organization of the United Nations, WHO, the United Nations Industrial Development Organization, the United Nations Institute for Training and Research and the Organisation for Economic Co-operation and Development (Participating Organizations), following recommendations made by the 1992 UN Conference on Environment and Development to strengthen cooperation and increase coordination in the field of chemical safety. The purpose of the IOMC is to promote coordination of the policies and activities pursued by the Participating Organizations, jointly or separately, to achieve the sound management of chemicals in relation to human health and the environment.

WHO Library Cataloguing-in-Publication Data

Nitrobenzene.

(Environmental health criteria ; 230)

1.Nitrobenzenes - toxicity 2.Nitrobenzenes - adverse effects 3.Environmental exposure 4.Risk assessment I.International Programme for Chemical Safety II.Series

ISBN 92 4 157230 0   (LC/NLM classification: QV 632)
ISSN 0250-863X

The Federal Ministry of the Environment, Nature Conservation, and Nuclear Safety, Germany, provided financial support for, and undertook the printing of, this publication.

CONTENTS

ENVIRONMENTAL HEALTH CRITERIA FOR NITROBENZENE

NOTE TO READERS OF THE CRITERIA MONOGRAPHS

Every effort has been made to present information in the criteria monographs as accurately as possible without unduly delaying their publication. In the interest of all users of the Environmental Health Criteria monographs, readers are requested to communicate any errors that may have occurred to the Director of the International Programme on Chemical Safety, World Health Organization, Geneva, Switzerland, in order that they may be included in corrigenda.

*   *   *

The WHO Environmental Health Criteria Programme is financially supported by the US Environmental Protection Agency, European Commission, German Federal Ministry of the Environment, Nature Conservation, and Nuclear Safety, and Japanese Ministry of Health, Labour and Welfare.

The Task Group meeting was arranged by the Fraunhofer Institute of Toxicology and Aerosol Research (now known as the Fraunhofer Institute of Inhalation Toxicology and Environmental Medicine), Germany.

Environmental Health Criteria

PREAMBLE

Objectives

In 1973 the WHO Environmental Health Criteria Programme was initiated with the following objectives:

(i)   to assess information on the relationship between exposure to environmental pollutants and human health, and to provide guidelines for setting exposure limits;
(ii)   to identify new or potential pollutants;
(iii)   to identify gaps in knowledge concerning the health effects of pollutants;
(iv)   to promote the harmonization of toxicological and epidemiological methods in order to have internationally comparable results.

The first Environmental Health Criteria (EHC) monograph, on mercury, was published in 1976, and since that time an ever-increasing number of assessments of chemicals and of physical effects have been produced. In addition, many EHC monographs have been devoted to evaluating toxicological methodology, e.g., for genetic, neurotoxic, teratogenic and nephrotoxic effects. Other publications have been concerned with epidemiological guidelines, evaluation of short-term tests for carcinogens, biomarkers, effects on the elderly and so forth.

Since its inauguration the EHC Programme has widened its scope, and the importance of environmental effects, in addition to health effects, has been increasingly emphasized in the total evaluation of chemicals.

The original impetus for the Programme came from World Health Assembly resolutions and the recommendations of the 1972 UN Conference on the Human Environment. Subsequently the work became an integral part of the International Programme on Chemical Safety (IPCS), a cooperative programme of UNEP, ILO and WHO. In this manner, with the strong support of the new partners, the importance of occupational health and environmental effects was fully recognized. The EHC monographs have become widely established, used and recognized throughout the world.

The recommendations of the 1992 UN Conference on Environment and Development and the subsequent establishment of the Intergovernmental Forum on Chemical Safety with the priorities for action in the six programme areas of Chapter 19, Agenda 21, all lend further weight to the need for EHC assessments of the risks of chemicals.

Scope

The criteria monographs are intended to provide critical reviews on the effect on human health and the environment of chemicals and of combinations of chemicals and physical and biological agents. As such, they include and review studies that are of direct relevance for the evaluation. However, they do not describe *every* study carried out. Worldwide data are used and are quoted from original studies, not from abstracts or reviews. Both published and unpublished reports are considered, and it is incumbent on the authors to assess all the articles cited in the references. Preference is always given to published data. Unpublished data are used only when relevant published data are absent or when they are pivotal to the risk assessment. A detailed policy statement is available that describes the procedures used for unpublished proprietary data so that this information can be used in the evaluation without compromising its confidential nature (WHO (1990) Revised Guidelines for the Preparation of Environmental Health Criteria Monographs. PCS/90.69, Geneva, World Health Organization).

In the evaluation of human health risks, sound human data, whenever available, are preferred to animal data. Animal and *in vitro* studies provide support and are used mainly to supply evidence missing from human studies. It is mandatory that research on human subjects is conducted in full accord with ethical principles, including the provisions of the Helsinki Declaration.

The EHC monographs are intended to assist national and international authorities in making risk assessments and subsequent risk management decisions. They represent a thorough evaluation of risks and are not, in any sense, recommendations for regulation or standard setting. These latter are the exclusive purview of national and regional governments.

Content

The layout of EHC monographs for chemicals is outlined below.

- Summary — a review of the salient facts and the risk evaluation of the chemical
- Identity — physical and chemical properties, analytical methods
- Sources of exposure
- Environmental transport, distribution and transformation
- Environmental levels and human exposure
- Kinetics and metabolism in laboratory animals and humans
- Effects on laboratory mammals and *in vitro* test systems
- Effects on humans
- Effects on other organisms in the laboratory and field
- Evaluation of human health risks and effects on the environment
- Conclusions and recommendations for protection of human health and the environment
- Further research
- Previous evaluations by international bodies, e.g., IARC, JECFA, JMPR

Selection of chemicals

Since the inception of the EHC Programme, the IPCS has organized meetings of scientists to establish lists of priority chemicals for subsequent evaluation. Such meetings have been held in Ispra, Italy, 1980; Oxford, United Kingdom, 1984; Berlin, Germany, 1987; and North Carolina, USA, 1995. The selection of chemicals has been based on the following criteria: the existence of scientific evidence that the substance presents a hazard to human health and/or the environment; the possible use, persistence, accumulation or degradation of the substance shows that there may be significant human or environmental exposure; the size and nature of populations at risk (both human and other species) and risks for environment; international concern, i.e., the substance is of major interest to several countries; adequate data on the hazards are available.

If an EHC monograph is proposed for a chemical not on the priority list, the IPCS Secretariat consults with the Cooperating Organizations and all the Participating Institutions before embarking on the preparation of the monograph.

Procedures

The order of procedures that result in the publication of an EHC monograph is shown in the flow chart on p. xv. A designated staff member of IPCS, responsible for the scientific quality of the document, serves as Responsible Officer (RO). The IPCS Editor is responsible for layout and language. The first draft, prepared by consultants or, more usually, staff from an IPCS Participating Institution, is based on extensive literature searches from reference databases such as Medline and Toxline.

The draft document, when received by the RO, may require an initial review by a small panel of experts to determine its scientific quality and objectivity. Once the RO finds the document acceptable as a first draft, it is distributed, in its unedited form, to well over 150 EHC contact points throughout the world who are asked to comment on its completeness and accuracy and, where necessary, provide additional material. The contact points, usually designated by governments, may be Participating Institutions, IPCS Focal Points or individual scientists known for their particular expertise. Generally some four months are allowed before the comments are considered by the RO and author(s). A second draft incorporating comments received and approved by the Director, IPCS, is then distributed to Task Group members, who carry out the peer review, at least six weeks before their meeting.

The Task Group members serve as individual scientists, not as representatives of any organization, government or industry. Their function is to evaluate the accuracy, significance and relevance of the information in the document and to assess the health and environmental risks from exposure to the chemical. A summary and recommendations for further research and improved safety aspects are also required. The composition of the Task Group is dictated by the range of expertise required for the subject of the meeting and by the need for a balanced geographical distribution.

EHC PREPARATION FLOW CHART

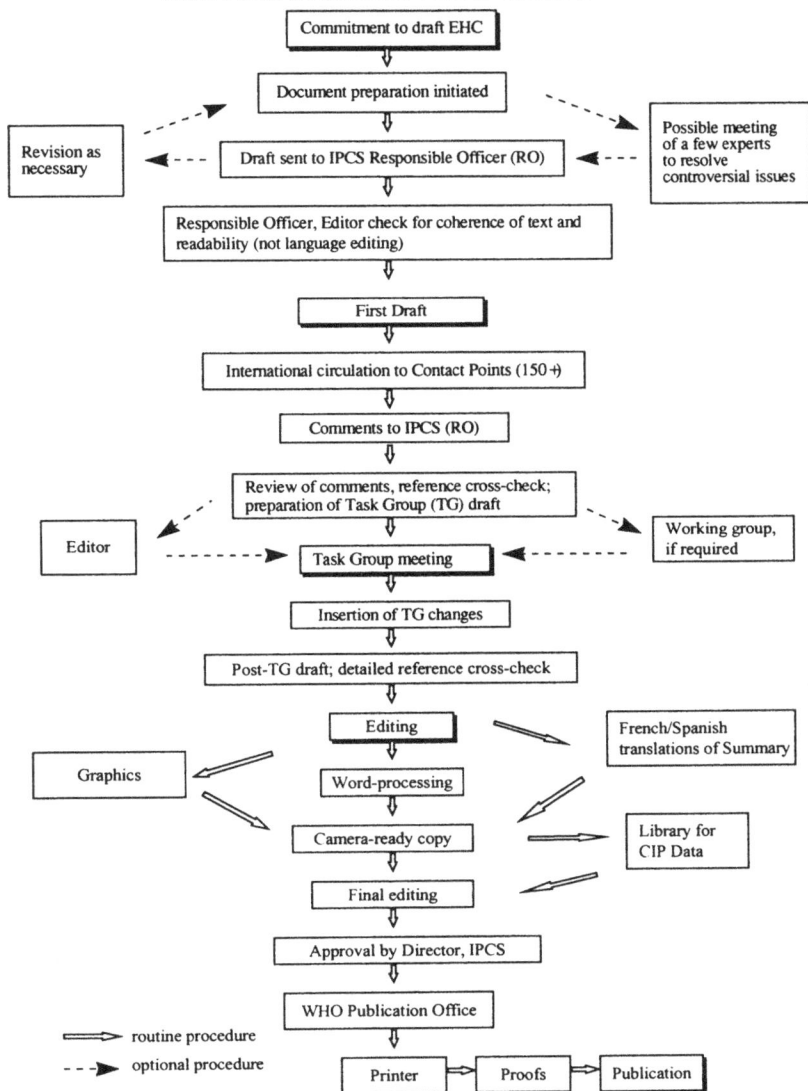

Commitment to draft EHC

⇩

Document preparation initiated

⇩

Revision as necessary ← - - - Draft sent to IPCS Responsible Officer (RO) ← - - - Possible meeting of a few experts to resolve controversial issues

⇩

Responsible Officer, Editor check for coherence of text and readability (not language editing)

⇩

First Draft

⇩

International circulation to Contact Points (150 +)

⇩

Comments to IPCS (RO)

⇩

Review of comments, reference cross-check; preparation of Task Group (TG) draft

⇩

Editor - - - - - → Task Group meeting ← - - - - - Working group, if required

⇩

Insertion of TG changes

⇩

Post-TG draft; detailed reference cross-check

⇩

Editing ⟹ French/Spanish translations of Summary

⇩

Graphics ⟸ Word-processing ⟹

⇩

Camera-ready copy ⟹ Library for CIP Data

⇩

Final editing

⇩

Approval by Director, IPCS

⇩

⟹ routine procedure

- - → optional procedure

WHO Publication Office

⇩

Printer ⇨ Proofs ⇨ Publication

The three cooperating organizations of the IPCS recognize the important role played by nongovernmental organizations. Representatives from relevant national and international associations may be invited to join the Task Group as observers. Although observers may provide a valuable contribution to the process, they can speak only at the invitation of the Chairperson. Observers do not participate in the final evaluation of the chemical; this is the sole responsibility of the Task Group members. When the Task Group considers it to be appropriate, it may meet *in camera*.

All individuals who as authors, consultants or advisers participate in the preparation of the EHC monograph must, in addition to serving in their personal capacity as scientists, inform the RO if at any time a conflict of interest, whether actual or potential, could be perceived in their work. They are required to sign a conflict of interest statement. Such a procedure ensures the transparency and probity of the process.

When the Task Group has completed its review and the RO is satisfied as to the scientific correctness and completeness of the document, it then goes for language editing, reference checking and preparation of camera-ready copy. After approval by the Director, IPCS, the monograph is submitted to the WHO Office of Publications for printing. At this time a copy of the final draft is sent to the Chairperson and Rapporteur of the Task Group to check for any errors.

It is accepted that the following criteria should initiate the updating of an EHC monograph: new data are available that would substantially change the evaluation; there is public concern for health or environmental effects of the agent because of greater exposure; an appreciable time period has elapsed since the last evaluation.

All Participating Institutions are informed, through the EHC progress report, of the authors and institutions proposed for the drafting of the documents. A comprehensive file of all comments received on drafts of each EHC monograph is maintained and is available on request. The Chairpersons of Task Groups are briefed before each meeting on their role and responsibility in ensuring that these rules are followed.

WHO TASK GROUP ON ENVIRONMENTAL HEALTH CRITERIA FOR NITROBENZENE

Members

Dr John Cocker, Health and Safety Laboratory, Sheffield, United Kingdom

Dr Les Davies, Office of Chemical Safety, Therapeutic Goods Administration, Australian Department of Health and Ageing, Canberra, Australia

Dr James W. Holder, National Center for Environmental Assessment, Office of Research and Development, US Environmental Protection Agency, Washington, DC, USA

Dr Jorma Mäki-Paakkanen, Division of Environmental Medicine, National Public Health Institute, Kuopio, Finland

Dr Inge Mangelsdorf, Fraunhofer Institute of Toxicology and Experimental Medicine, Hanover, Germany *(Chairperson)*

Ms Fatoumata Jallow Ndoye, National Environment Agency, Banjul, Gambia

Dr Sumol Pavittranon, Toxicology and Environmental Laboratory, Ministry of Public Health, Nonthaburi, Thailand

Dr Vesa Riihimäki, Finnish Institute of Occupational Health, Helsinki, Finland *(Vice-Chairperson)*

Dr Gilbert Schönfelder, Institute for Clinical Pharmacology and Toxicology, University Hospital Benjamin Franklin, Berlin, Germany

Dr Jenny Stauber, CSIRO Energy Technology, Bangor, NSW, Australia *(Rapporteur)*

Secretariat

Ms Vida Gyamerah, International Programme on Chemical Safety, World Health Organization, Geneva, Switzerland

Mr Yoshikazu Hayashi, International Programme on Chemical Safety, World Health Organization, Geneva, Switzerland

Dr J. Kielhorn, Fraunhofer Institute of Toxicology and Experimental Medicine, Hanover, Germany

Observer

Dr Friedhelm Pilger, Bayer AG, Leverkusen, Germany

ENVIRONMENTAL HEALTH CRITERIA FOR NITROBENZENE

A WHO Task Group on Environmental Health Criteria for Nitro-benzene met at the Fraunhofer Institute of Toxicology and Experimental Medicine, Hanover, Germany, from 19 to 23 February 2001. Mr Y. Hayashi, Programme for the Promotion of Chemical Safety, WHO, opened the meeting and welcomed the participants on behalf of the IPCS and its three cooperative organizations (UNEP/ILO/WHO). The Task Group reviewed and revised the draft monograph and made an evaluation of the risks for human health and the environment from exposure to nitrobenzene.

The first draft was prepared by Dr Les Davies, Therapeutic Goods Administration, Commonwealth Department of Health and Family Services, Australia. The second draft was also prepared by the same author, who incorporated comments received following the circulation of the first draft to the IPCS contact points for Environmental Health Criteria monographs.

Mr Y. Hayashi and Mr. T. Ehara were responsible for the overall scientific content of the monograph, and Dr A. Aitio was responsible for coordinating the technical editing of the monograph.

The efforts of all who helped in the preparation and finalization of the monograph are gratefully acknowledged.

* * *

Financial support for this Task Group was provided by the Government of Japan and the Government of Germany as part of their contributions to the IPCS.

ACRONYMS AND ABBREVIATIONS

AMP	adenosine 5'-monophosphate
BCF	bioconcentration factor
BMD	benchmark dose
BOD	biochemical oxygen demand
BOD_5	5-day biochemical oxygen demand
cAMP	cyclic adenosine 5'-monophosphate
CAS	Chemical Abstracts Service
CIIT	Chemical Industry Institute of Toxicology
COD	chemical oxygen demand
DMPO	5,5-dimethyl-1-pyrroline-N-oxide
DNA	deoxyribonucleic acid
DOC	dissolved organic carbon
EC_{50}	median effective concentration
EDTA	ethylenediaminetetraacetic acid
EHC	Environmental Health Criteria monograph
EPA	Environmental Protection Agency (USA)
ESR	electron spin resonance
EU	European Union
FAO	Food and Agriculture Organization of the United Nations
FID	flame ionization detector
FSH	follicle stimulating hormone
FTIR	Fourier transform infrared spectrometry
GC	gas chromatography
GSH	glutathione
GSSG	oxidized glutathione dimer
Hb	haemoglobin
HC_5	hazardous concentration to protect 95% of species with an associated confidence level
HPLC	high-performance liquid chromatography
HRGC	high-resolution gas chromatography
IARC	International Agency for Research on Cancer
IC_{50}	median inhibitory concentration
ILO	International Labour Organization
IPCS	International Programme on Chemical Safety
IUPAC	International Union of Pure and Applied Chemistry
JECFA	Joint Expert Committee on Food Additives and Contaminants

JMPR	Joint FAO/WHO Meeting on Pesticide Residues
K_d	soil sorption coefficient
K_{oc}	organic carbon/water partition coefficient
K_{om}	organic matter/water partition coefficient
K_{ow}	octanol/water partition coefficient
K_{sed}	soil/sediment sorption coefficient
LC_{50}	median lethal concentration
LMS	linearized multistage
LOAEL	lowest-observed-adverse-effect level
LOEC	lowest-observed-effect concentration
LOEL	lowest-observed-effect level
MDT	maximum dose tested
MITI	Ministry of International Trade and Industry (Japan)
MS	mass spectrometry
MTD	maximum tolerated dose
NAD	nicotinamide adenine dinucleotide
NADP	nicotinamide adenine dinucleotide phosphate
NADPH	nicotinamide adenine dinucleotide phosphate, reduced form
NIOSH	National Institute for Occupational Safety and Health (USA)
NOAEL	no-observed-adverse-effect level
NOEC	no-observed-effect concentration
NOEL	no-observed-effect level
NTP	National Toxicology Program (USA)
OECD	Organisation for Economic Co-operation and Development
ppb	part per billion (one billion = 1 000 000 000)
ppm	part per million
ppt	part per trillion
RO	Responsible Officer
RTECS	Registry of Toxic Effects of Chemical Substances
SI	International System of Units (Système international d'unités)
STORET	STOrage and RETrieval (US EPA database)
$t_{1/2}$	half-life
TLV	threshold limit value
TOC	total organic carbon

UCL	upper confidence limit
UN	United Nations
UNEP	United Nations Environment Programme
UV	ultraviolet
VSD	virtually safe dose
WHO	World Health Organization

1. SUMMARY

1.1 Identity, physical and chemical properties, and analytical methods

Nitrobenzene is a colourless to pale yellow oily liquid with an odour resembling that of bitter almonds or "shoe polish." It has a melting point of 5.7 °C and a boiling point of 211 °C. Its vapour pressure is 20 Pa at 20 °C, and its solubility in water is 1900 mg/litre at 20 °C. It represents a fire hazard, with a flash point (closed cup method) of 88 °C and an explosive limit (lower) of 1.8% by volume in air. Its log octanol/water partition coefficient is 1.85.

A range of analytical methods is available for the quantification of nitrobenzene in air, water and soil samples. Methods for the monitoring of workers systemically exposed to nitrobenzene are also available. A reversed-phase high-performance liquid chromatographic method for the determination of urinary metabolites, including p-nitrophenol (also a urinary metabolite of the organophosphorus insecticides parathion and parathion-methyl), appears suitable. Methods are also available for the determination of aniline released from haemoglobin adducts and for the determination of methaemoglobin, which is produced by nitrobenzene metabolites.

1.2 Sources of human and environmental exposure

Nitrobenzene does not occur naturally. It is a synthetic compound, more than 95% of which is used in the production of aniline, a major chemical intermediate that is used in the manufacture of polyurethanes; nitrobenzene is also used as a solvent in petroleum refining, as a solvent in the manufacture of cellulose ethers and acetates, in the manufacture of dinitrobenzenes and dichloroanilines, and in the synthesis of other organic compounds, including acetaminophen.

Early in the 20th century, nitrobenzene had some use as a food additive (substitute for almond essence) as well as extensive use as a solvent in various proprietary products, including boot polish, inks (including inks used for stamping freshly laundered hospital baby diapers) and several disinfectants, so there was a significant potential for public exposure at that time.

From available records, it is apparent that there has been a significant increase in annual production of nitrobenzene over the past 30–40 years. Most is retained in closed systems for use in further synthesis, particularly of aniline, but also of substituted nitrobenzenes and anilines. Losses during production of nitrobenzene are likely to be low; when nitrobenzene is used as a solvent, however, emissions may be higher. Nitrobenzene has been shown to be emitted from sewage sludge incineration units and has been measured in air at hazardous waste landfills.

Nitrobenzene can be formed from the atmospheric reaction of benzene in the presence of nitrogen oxides, although this source has not been quantified. Aniline has been reported to be slowly oxidized by ozone to nitrobenzene.

1.3 Environmental transport, distribution and transformation

Nitrobenzene can undergo degradation by both photolysis and microbial biodegradation.

The physical properties of nitrobenzene suggest that transfer from water to air will be significant, although not rapid. Photodegradation of nitrobenzene in air and water is slow. From direct photolysis experiments in air, lifetimes of <1 day were determined, whereas the calculated half-lives for the reaction with hydroxyl radicals were in the range between 19 and 223 days. With ozone, the reaction proceeds even more slowly. Experiments in a smog chamber with a propylene/butane/nitrogen dioxide mixture gave an estimated lifetime for nitrobenzene between 4 and 5 days. In water bodies, direct photolysis appears to be the degradation pathway that proceeds most rapidly (half-lives between 2.5 and 6 days), whereas indirect photolysis (photo-oxidation with hydroxyl radicals, hydrogen atoms or hydrated electrons, sensitization with humic acids) plays a minor role (calculated half-lives between 125 days and 13 years for the reaction with hydroxyl radicals, depending on sensitizer concentration).

2

Due to its moderate water solubility and relatively low vapour pressure, it might be expected that nitrobenzene would be washed out of the atmosphere by rain to some extent; however, in field experiments, it appeared that washout by rainfall (either through solution in raindrops or by removal of nitrobenzene sorbed onto particulates) and dryfall of particulates was negligible. Because of its vapour density (4.1–4.25 times that of air), removal processes from the atmosphere may include settling of vapours.

Actual data on evaporation of nitrobenzene from water bodies appear to be somewhat conflicting, with a computer model predicting volatilization half-lives of 12 days (river) to 68 days (eutrophic lake). The shortest estimate cited in the literature was 1 day (from river water); in another study of experimental microcosms, simulating land application of wastewater, nitrobenzene was reported not to volatilize but to be totally degraded.

Degradation studies suggest that nitrobenzene is degraded in sewage treatment plants by aerobic processes, with slower degradation under anaerobic conditions. Nitrobenzene may not necessarily be completely degraded if it is present at high concentrations in wastewater. High concentrations may also inhibit the biodegradation of other wastes. Biodegradation of nitrobenzene depends mainly on the acclimation of the microbial population. Degradation by non-acclimated inocula is generally very slow to negligible and proceeds only after extended acclimation periods. Acclimated microorganisms, particularly from industrial wastewater treatment plants, however, showed complete elimination of nitrobenzene within a few days. Degradation was generally found to be increased in the presence of other easily degradable substrates. Adaptation of the microflora and additional substrates also seem to be the limiting factors for the decomposition of nitrobenzene in soil. Degradation of nitrobenzene under anaerobic conditions has been shown to be very slow, even after extended acclimation periods.

The measured bioconcentration factors for nitrobenzene in a number of organisms indicate minimal potential for bioaccumulation, and nitrobenzene is not biomagnified through the food-chain. Nitrobenzene may be taken up by plants; in available studies, however, it appeared to be associated with roots, and very little was associated with other parts of the plant. In a simulated "farm pond" aquatic ecosystem, nitrobenzene remained mainly in the water and was neither stored nor ecologically magnified in water fleas, mosquito larvae, snails, algae, miscellaneous plankton or fish.

1.4 Environmental levels and human exposure

Concentrations of nitrobenzene in environmental samples such as surface water, groundwater and air are generally low.

Some measured levels in air in US cities in the early 1980s ranged between <0.05 and 2.1 $\mu g/m^3$ (<0.01 and 0.41 ppb) (arithmetic means). Data reported by the US Environmental Protection Agency in 1985 indicated that less than 25% of air samples in the USA were positive, with a median concentration of about 0.05 $\mu g/m^3$ (0.01 ppb); in urban areas, mean levels were generally less than 1 $\mu g/m^3$ (0.2 ppb), with slightly higher levels in industrial areas (mean 2.0 $\mu g/m^3$ [0.40 ppb]). Of 49 air samples measured in Japan in 1991, 42 had a detectable level, measured as 0.0022–0.16 $\mu g/m^3$. Levels over urban areas and waste disposal sites were significantly lower (or undetectable) in winter than in summer.

Data on nitrobenzene levels in surface water appear to be more extensive than data on levels in air. While levels are variable depending on location and season, generally low levels (around 0.1–1 $\mu g/litre$) have been measured. One of the highest levels reported was 67 $\mu g/litre$, in the river Danube, Yugoslavia, in 1990. However, nitrobenzene was not detected in any surface water samples collected near a large number of hazardous waste sites in the USA (reported in 1988). Based on limited data, it appears that there may be greater potential for contamination of groundwater than of surface water; several sites measured in the USA in the late 1980s had levels of 210–250 and 1400 $\mu g/litre$ (with much higher levels at a coal gasification site). Nitrobenzene has been reported in studies conducted in the 1970s and 1980s on drinking-water in the USA and the United Kingdom, albeit in only a small proportion of samples, but was not detected in 30 Canadian samples (1982 report).

No data on nitrobenzene occurrence in food were located, although Japanese studies conducted in 1991 detected it in a small proportion (4 of 147) of fish samples. It was not detected in a large range of sampled biota in a 1985 US study.

The general population can be exposed to variable concentrations of nitrobenzene in air and possibly drinking-water. There is also potential exposure from consumer products, but accurate information is lacking. In studies conducted in the state of New Jersey on the eastern coast of the USA (warm to hot summers and cold to very cold winters), urban areas had higher levels in summer than in winter due to both the formation of nitrobenzene by nitration of benzene (from petrol) and the higher volatility of nitrobenzene during the warmer months; ambient air exposure in the winter may be negligible. Based on air studies and on estimates of releases during manufacture, only populations in the vicinity of manufacturing activities (i.e., producers and industrial consumers of nitrobenzene for subsequent synthesis) and petroleum refining plants are likely to have any significant exposure to nitrobenzene. However, people living in and around abandoned hazardous waste sites may also have the potential for higher exposure, due to possible groundwater and soil contamination and uptake of nitrobenzene by plants.

Occupational exposure levels should be less than the widely adopted airborne exposure limit of 5 mg/m^3 (1 ppm). Based on available data, it appears that nitrobenzene is well absorbed dermally, both in vapour form and as a liquid; therefore, dermal exposure may be significant, but data are lacking.

1.5   Kinetics and metabolism

Nitrobenzene is a volatile liquid that can readily gain access to the body by inhalation and skin penetration of the vapour, as well as by ingestion and dermal absorption of the liquid. Nitrobenzene activation in rats to methaemoglobin-forming metabolites appears to be mediated to a significant degree by intestinal microflora. In test animals, the major part of nitrobenzene (about 80% of the dose) is metabolized and eliminated within 3 days. The remainder is eliminated only slowly. The slow compartment is likely due to erythrocyte recycling of nitrobenzene redox forms and glutathione conjugates. Covalent binding, presumably to sulfhydryl groups of haemoglobin, was demonstrated.

In rodents and rabbits, *p*-nitrophenol and *p*-aminophenol are major urinary metabolites. In humans, part of the absorbed dose is excreted into the urine; 10–20% of the dose is excreted as *p*-nitrophenol (which thus may be used for biological monitoring). The half-

5

times of elimination for *p*-nitrophenol are estimated to be about 5 h (initial phase) and >20 h (late phase). The urinary metabolite *p*-aminophenol is significant only at higher doses.

1.6 Effects on laboratory mammals and *in vitro* test systems

Nitrobenzene causes toxicity in multiple organs by all routes of exposure. Methaemoglobinaemia results from oral, dermal, subcutaneous and inhalational nitrobenzene exposure in mice and rats, with consequent haemolytic anaemia, splenic congestion and liver, bone marrow and spleen haematopoiesis.

Splenic capsular lesions were seen in rats by both gavage (at doses as low as 18.75 mg/kg of body weight per day) and dermal (at 100 mg/kg of body weight per day and above) routes of administration. Similar splenic lesions have previously been observed with aniline-based dyes, some of which produced splenic sarcomas in chronic carcinogenicity studies in rats. Effects on the liver were noted in mice and rats after both gavage and dermal administration of nitrobenzene, with centrilobular hepatocyte necrosis, hepatocellular nucleolar enlargement, severe hydropic degeneration and pigment accumulation in Kupffer cells reported. Increased vacuolation of the X-zone of the adrenal gland was noted in female mice after oral and dermal dosing.

In subchronic oral and dermal studies in mice and rats, central nervous system lesions in the cerebellum and brain stem were life-threatening. These lesions, including petechial haemorrhages, may be direct toxic effects or mediated by vascular effects of hypoxia or hepatic toxicity. Depending on the dose, these neurotoxic effects were grossly apparent as ataxia, head-tilt and arching, loss of righting reflex, tremors, coma and convulsions.

Other target organs included kidney (increased weight, glomerular and tubular epithelial swelling, pigmentation of tubular epithelial cells), nasal epithelium (glandularization of the respiratory epithelium, pigment deposition in and degeneration of olfactory epithelium), thyroid (follicular cell hyperplasia), thymus (involution) and pancreas (mononuclear cell infiltration), while lung pathology (emphysema, atelectasis and bronchiolization of alveolar cell walls) was reported in rabbits.

The potential carcinogenicity and toxicity of inhaled nitrobenzene were evaluated following long-term exposure (505 days) of male and female B6C3F$_1$ mice, male and female Fischer-344 rats and male Sprague-Dawley rats. Survival was not adversely affected at the concentrations tested (up to 260 mg/m^3 [50 ppm] for mice; up to 130 mg/m^3 [25 ppm] for rats), but inhaled nitrobenzene was toxic and carcinogenic in both species and both rat strains, inducing a spectrum of benign and malignant (lung, thyroid, mammary gland, liver, kidney) neoplasias.

Nitrobenzene was non-genotoxic in bacteria and mammalian cells *in vitro* and in mammalian cells *in vivo*. Studies reported included DNA damage and repair assays, gene mutation assays, chromosomal effects assays and cell transformation assays.

Numerous studies have confirmed that nitrobenzene is a testicular toxicant, with the most sensitive spermatic end-points being sperm count and motility, followed by progressive motility, viability, presence of abnormal sperm and, finally, the fertility index.

In a two-generation reproductive toxicity study in Sprague-Dawley rats by the inhalational route, nitrobenzene at 200 mg/m^3 (40 ppm), but not at 5 or 51 mg/m^3 (1 or 10 ppm), caused a large decrease in the fertility index of F$_0$ and F$_1$ generations, associated with male reproductive system toxicity; this decreased fertility was partially reversible, when the F$_1$ generation from the 200 mg/m^3 group was mated with virgin untreated females after a 9-week recovery period. However, in an oral dosing study in the same rat strain (20–100 mg/kg of body weight from 14 days premating to day 4 of lactation), while pup body weight was lowered and postnatal loss was increased, nitrobenzene was without effect on reproductive parameters. The lack of effect on fertility in this study was due to the short premating dosage interval and the fact that rats produce sperm in very large excess. Impaired male fertility with significant testicular atrophy was seen in mice and rats; effects in mice were apparent at gavage doses of 300 mg/kg of body weight per day and dermal doses of 800 mg/kg of body weight per day and in rats at gavage doses of 75 mg/kg of body weight per day and dermal doses of 400 mg/kg of body weight per day. Testicular toxicity was seen as desquamation of the seminiferous epithelium, the appearance of multinucleated giant cells, gross atrophy and prolonged aspermia. Nitrobenzene has direct effects on the testis, shown by *in vivo* and *in vitro* studies. Spermatogenesis is affected, with exfoliation of predominantly viable germ cells and degenerating Sertoli cells. The main histopathological effects are degenerated spermatocytes.

In general, maternal reproductive organs were not affected, except for one study where uterine atrophy was seen in mice after a dermal dose of 800 mg/kg of body weight per day.

Developmental toxicity studies in rats and rabbits indicated that inhalation exposure to nitrobenzene did not result in fetotoxic, embryotoxic or teratogenic effects at concentrations sufficient to produce maternal toxicity. At the highest concentration tested in these studies (530 mg/m^3 [104 ppm] in a rabbit study), the mean numbers of resorption sites and percentage of resorptions/implants were higher in this group than in concurrent controls, but were within the historical control range; maternal effects (i.e., increased methaemoglobin levels and increased liver weights) were noted from 210 mg/m^3 (41 ppm).

In a study on the immunotoxicity of nitrobenzene in B6C3F$_1$ mice, nitrobenzene caused increased cellularity of the spleen, a degree of immunosuppression (diminished IgM response to sheep red blood cells) and bone marrow stimulation. Host resistance to microbial or viral infection was not markedly affected by nitrobenzene, although there was a trend towards increased susceptibility in cases in which T-cell function contributed to host defence.

1.7 Effects on humans

Nitrobenzene is toxic to humans by inhalational, dermal and oral exposure. The main systemic effect associated with human exposure to nitrobenzene is methaemoglobinaemia.

Numerous accidental poisonings and deaths in humans from ingestion of nitrobenzene have been reported. In cases of oral ingestion or in which the patients were apparently near death due to severe methaemoglobinaemia, termination of exposure and prompt medical intervention resulted in gradual improvement and recovery. Although human exposure to sufficiently high quantities of nitrobenzene can be lethal via any route of exposure, it is considered unlikely that levels of exposure high enough to cause death would occur except in cases of industrial accidents or suicides.

The spleen is likely to be a target organ during human exposure to nitrobenzene; in a woman occupationally exposed to nitrobenzene in paint (mainly by inhalation), the spleen was tender and enlarged.

Liver effects, including hepatic enlargement and tenderness and altered serum chemistries, have been reported in a woman inhalationally exposed to nitrobenzene.

Neurotoxic symptoms reported in humans after inhalation exposure to nitrobenzene have included headache, confusion, vertigo and nausea. Effects in orally exposed persons have also included those symptoms, as well as apnoea and coma.

1.8   Effects on organisms in the environment

Nitrobenzene appears to be toxic to bacteria and may adversely affect sewage treatment facilities if present in high concentrations in influent. The lowest toxic concentration reported for microorganisms is for the bacterium *Nitrosomonas*, with an EC_{50} of 0.92 mg/litre based upon the inhibition of ammonia consumption. Other reported values are a 72-h no-observed-effect concentration (NOEC) of 1.9 mg/litre for the protozoan *Entosiphon sulcatum* and an 8-day lowest-observed-effect concentration (LOEC) of 1.9 mg/litre for the blue-green alga *Microcystis aeruginosa*.

For freshwater invertebrates, acute toxicity (24- to 48-h LC_{50} values) ranged from 24 mg/litre for the water flea (*Daphnia magna*) to 140 mg/litre for the snail (*Lymnaea stagnalis*). For marine invertebrates, the lowest acute toxicity value reported was a 96-h LC_{50} of 6.7 mg/litre for the mysid shrimp (*Mysidopsis bahia*). The lowest chronic test value reported was a 20-day NOEC of 1.9 mg/litre for *Daphnia magna*, with an EC_{50}, based on reproduction, of 10 mg/litre.

Freshwater fish showed similar low sensitivity to nitrobenzene. The 96-h LC_{50} values ranged from 24 mg/litre for the medaka (*Oryzias latipes*) to142 mg/litre for the guppy (*Poecilia reticulata*). There was no effect on mortality or behaviour of medaka at 7.6 mg/litre over an 18-day exposure.

1.9 Hazard and risk evaluation

Methaemoglobinaemia and subsequent haematological and splenic changes have been observed in exposed humans, but the data do not allow quantification of the exposure–response relationship. In rodents, methaemoglobinaemia, haematological effects, testicular effects and, in the inhalation studies, effects on the respiratory system were found at the lowest doses tested. Methaemoglobinaemia, bilateral epididymal hypospermia and bilateral testicular atrophy were observed at the lowest exposure level studied, 5 mg/m³ (1 ppm), in rats. In mice, there was a dose-related increase in the incidence of bronchiolization of alveolar walls and alveolar/bronchial hyperplasia at the lowest dose tested of 26 mg/m³ (5 ppm). Carcinogenic response was observed after exposure to nitrobenzene in rats and mice: mammary adenocarcinomas were observed in female B6C3F$_1$ mice, and liver carcinomas and thyroid follicular cell adenocarcinomas were seen in male Fischer-344 rats. Benign tumours were observed in five organs. Studies on geno-toxicity have usually given negative results.

Although several metabolic products of nitrobenzene are candidates for cancer causality, the mechanism of carcinogenic action is not known. Because of the likely commonality of redox mechanisms in test animals and humans, it is hypothesized that nitrobenzene may cause cancer in humans by any route of exposure.

Exposure of the general population to nitrobenzene from air or drinking-water is likely to be very low. Although no no-observed-adverse-effect level (NOAEL) could be derived from any of the toxicological studies, there is a seemingly low risk for non-neoplastic effects. If exposure values are low enough to avoid non-neoplastic effects, it is expected that carcinogenic effects will not occur.

Acute poisonings by nitrobenzene in consumer products have occurred frequently in the past. Significant human exposure is pos-sible, due to the moderate vapour pressure of nitrobenzene and extensive skin absorption. Furthermore, the relatively pleasant almond smell of nitrobenzene may not discourage people from consuming food or water contaminated with it. Infants are especially susceptible to the effects of nitrobenzene.

There is limited information on exposure in the workplace. In one workplace study, exposure concentrations were of the same order of magnitude as the lowest-observed-adverse-effect levels (LOAELs) in a long-term inhalation study. Therefore, there is significant concern for the health of workers exposed to nitrobenzene.

Nitrobenzene shows little tendency to bioaccumulate and appears to undergo both aerobic and anaerobic biotransformation. For terrestrial systems, the levels of concern reported in laboratory tests are unlikely to occur in the natural environment, except possibly in areas close to nitrobenzene production and use and areas contaminated by spillage.

Using the available acute toxicity data and a statistical distribution method, together with an acute:chronic toxicity ratio derived from data on crustaceans, the concentration limit for nitrobenzene to protect 95% of freshwater species with 50% confidence may be estimated to be 200 µg/litre. Nitrobenzene is thus unlikely to pose an environmental hazard to aquatic species at levels typically reported in surface waters, around 0.1–1 µg/litre. Even at highest reported concentrations (67 µg/litre), nitrobenzene is unlikely to be of concern to freshwater species.

There is not enough information to derive a guideline value for marine organisms.

2. IDENTITY, PHYSICAL AND CHEMICAL PROPERTIES, AND ANALYTICAL METHODS

2.1 Identity

Common name: nitrobenzene
Chemical formula: $C_6H_5NO_2$
Chemical structure:

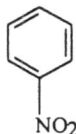

Relative molecular mass: 123.11
CAS name: nitrobenzene
IUPAC name: nitrobenzene
CAS registry number: 98-95-3
NIOSH RTECS DA6475000
Synonyms: nitrobenzol, mononitrobenzol, MNB, C.I. solvent black 6, essence of mirbane, essence of myrbane, mirbane oil, oil of mirbane, oil of myrbane, nigrosine spirit soluble B
Trade name: Caswell No. 600

2.2 Physical and chemical properties

Nitrobenzene is a colourless to pale yellow oily liquid that presents a fire hazard. Its odour resembles that of bitter almonds or "shoe polish," with reported odour thresholds of 0.092 mg/m³ (0.018 ppm) (Amoore & Hautala, 1983) and 0.03 mg/m³ (0.005 ppm) (Manufacturing Chemists Association, 1968). The odour threshold in water has been reported as 0.11 mg/litre (Amoore & Hautala, 1983) and 0.03 mg/litre (US EPA, 1980). Some chemical and physical properties of nitrobenzene are given in Table 1.

Table 1. Some physical and chemical properties of nitrobenzene[a]

Property	Value
Specific gravity	1.2037 at 20 °C
	1.205 at 28 °C
Melting point	5.7 °C
Boiling point	211 °C
Vapour pressure	20 Pa (0.15 mmHg) at 20 °C
	38 Pa (0.284 mmHg) at 25 °C
	47 Pa (0.35 mmHg) at 30 °C
Vapour density	4.25[b]
Flash point (closed cup)	88 °C
Explosive limit (lower)	1.8% by volume in air
Solubility in water	1900 mg/litre at 20 °C
	2090[c] mg/litre at 25 °C
Solubility in organic solvents	Freely soluble in ethanol, benzene, acetone, ether and oils
Octanol/water partition coefficient (log K_{ow})	1.85 (1.6–2.0)
Organic carbon/water partition coefficient (log K_{oc})	1.56
Henry's law constant	
(measured)	2.4[d] Pa·m³/mol (20 °C)
	0.868[e] Pa·m³/mol (25 °C)
(calculated)	2.47[f] (dimensionless)
	1.3[g] Pa·m³/mol

[a] From ATSDR (1990); Howard et al. (1990); ACGIH (1991); BUA (1994).
[b] From Verschueren (1983).
[c] From Banerjee et al. (1980).
[d] From Warner et al. (1987).
[e] From Altschuh et al. (1999).
[f] From Howard et al. (1990), using a vapour pressure of 36 Pa (0.27 mmHg) and a water solubility of 2000 mg/litre.
[g] From Enfield et al. (1986).

2.3 Conversion factors in air

The conversion factors[1] for nitrobenzene in air (at 20 °C and 101.3 kPa) are as follows:

1 ppm = 5.12 mg/m^3
1 mg/m^3 = 0.195 ppm

2.4 Analytical methods

Albrecht & Neumann (1985) discussed some of the difficulties of analysing nitrobenzene and one of its metabolites, aniline, in animals. Excretion of the parent compound or some metabolites in urine has been determined, but there are practical and methodological problems (Albrecht & Neumann, 1985). Acute poisoning by nitrobenzene has been monitored by measuring levels of methaemoglobin, which is produced by nitrobenzene metabolites. However, many toxicants produce methaemoglobin, and this analysis is certainly not a satisfactory method for monitoring nitrobenzene in animals.

2.4.1 Sampling and pretreatment

Sampling on Tenax™ of different mesh sizes followed by thermal desorption and gas chromatography (GC) was evaluated as a simple method for the determination of nitrobenzene in workplace air (Patil & Lonkar, 1992). An alternative sampling technique in place of pump sampling was developed, utilizing a 10-litre water siphon bottle. Quantitative recoveries were obtained in the mass range 0.04–10 μg. Air humidity had no effect on recovery. The charged tubes could be stored at room temperature for 5 days with no change in recovery. The particle size of Tenax TA had no significant effect on adsorption or desorption (Patil & Lonkar, 1992).

---

[1] In keeping with WHO policy, which is to provide measurements in SI units, all concentrations of gaseous chemicals in air are given in SI units. Where the original study or source document has provided concentrations in SI units, these will be cited here. Where the original study or source document has provided concentrations in volumetric units, conversions will be done using the conversion factors given here, assuming a temperature of 20 °C and a pressure of 101.3 kPa. Conversions are to no more than two significant digits.

The US National Institute for Occupational Safety and Health (NIOSH) has developed an approach to estimating worker exposure to *o*-toluidine, aniline and nitrobenzene using a combination of surface wipe, dermal badge and air sampling (Pendergrass, 1994). The greatest recoveries of airborne nitrobenzene were from large-capacity silica gel sorbent tubes, and the limit of detection (1 µg) was claimed to be approximately 10 times more sensitive than those for previous NIOSH methods. In this study, recoveries of airborne nitrobenzene under different levels of humidity were investigated. Surface wipes (cotton gauze pads, 100 cm^2), dermal badges (cotton pouches filled with 0.75–80 g 8- to 20-mesh silica gel) and sorbent tubes were best desorbed with absolute ethanol in an ultrasonic bath.

2.4.2 Analysis

2.4.2.1 Environmental monitoring

Nitrobenzene has been determined in environmental samples (air, water, soil and solid waste samples) by GC analysis following collection and extraction with an organic solvent (US EPA, 1982a, 1982b; NIOSH, 1984); flame ionization or mass spectrometry (MS) may be used for detection.

Piotrowski (1967) used a colorimetric method to determine nitrobenzene levels in air, following absorption of the compound into 10 ml ice-cold water. Passing 5 litres of air through 10 ml of water resulted in a sensitivity of 3 µg, with a coefficient of deviation of 5%.

A trochoidal electron monochromator was interfaced to a mass spectrometer to perform electron capture negative ion mass spectrometric analyses of environmentally relevant chemicals; multiple resonance states resulting in stable radical anions were easily observed for nitrobenzene, although the study did not assess the sensitivity of the system (Laramée et al., 1992).

Analytical methods for the determination of nitrobenzene in environmental samples are given in Table 2.

2.4.2.2 Biological monitoring

Information about urinary metabolites of nitrobenzene is given in sections 6.3 and 6.7.

Table 2. Analytical methods for determining nitrobenzene in environmental samples

Sample matrix	Sample preparation	Analysis[a]	Detection limit	Accuracy (%)	Reference
Air at landfill sites	Adsorption on Tenax-GC cartridges, thermal desorption	HRGC/FID	0.3 µg/m³	No data	Harkov et al., 1985
Air	Adsorption on silica gel, extraction with methanol	GC/FID	0.02 mg/sample	No data	NIOSH, 1984
Air	Adsorption on silica gel, extraction with methanol	GC/FID	0.5 mg/m³	No data	NIOSH, 1977
Wastewater	Direct injection of aqueous sample	GC/FID	No data	No data	Patil & Shinde, 1988
Wastewater	Extract with dichloromethane, exchange to hexane, concentrate	GC/FID	3.6 µg/litre	71 ± 5.9[b]	US EPA, 1982a
Water	Extract with dichloromethane at pH 11 and 2, concentrate	GC/MS	1.9 µg/litre	71 ± 31[b]	US EPA, 1982b
Soil and solid waste	Extract from sample, clean up	GC/FID	137 mg/kg[c]	25.7–100[b]	US EPA, 1986b
Soil and solid waste	Extract from sample, clean up	GC/MS	19 mg/kg[c]	No data	US EPA, 1986c
Soil and solid waste	Extract from sample, clean up	GC/FID	660 µg/kg[d]	54–158[b]	US EPA, 1986d
Soil and solid waste	Extract from sample, clean up	HRGC/FTIR	12.5 µg/litre[e]	No data	US EPA, 1986a

Table 2 (Contd).

Sample matrix	Sample preparation	Analysis[a]	Detection limit	Accuracy (%)	Reference
Workplace air	Adsorption on Tenax-GC cartridges, thermal desorption	GC/FID	<0.04 µg/sample	No data	Patil & Lonkar, 1992
Workplace air	Adsorption on silica gel, extraction with absolute ethanol	GC/FID	1 µg	4.3[f]	Pendergrass, 1994

[a] FID = flame ionization detector; FTIR = Fourier transform infrared spectrometry; GC = gas chromatography; HRGC = high-resolution gas chromatography; MS = mass spectrometry.
[b] Relative recovery, percent, ± standard deviation.
[c] Approximate detection limit in high-level soil and sludges.
[d] Approximate detection limit in low-level soil and sediments.
[e] Detection limit in water. Detection limit in solids and wastes is several orders of magnitude higher.
[f] Relative standard deviation.

1) Urinary p-nitrophenol

The determination of p-nitrophenol in urine samples collected at the end of shift at the end of a workweek offers a simple, non-invasive approach to biological monitoring (see section 6.7). The American Conference of Governmental Industrial Hygienists has proposed a biological exposure index of 5 mg p-nitrophenol/g creatinine (ACGIH, 2000).

A reversed-phase high-performance liquid chromatographic (HPLC) method was described for the simultaneous determination of urinary metabolites of several aromatic chemicals, including p-nitrophenol (formed from nitrobenzene). The proposed method appears suitable for the routine monitoring of workers exposed to these chemicals (Harmer et al., 1989; Astier, 1992).

Bader et al. (1998) used two GC/MS-selected ion monitoring methods for the determination of nitroaromatic compounds, including nitrobenzene, in urine samples. Analytes were detectable in the lowest microgram per litre range, and both methods were useful for screening occupationally or environmentally exposed people.

The p-nitrophenol metabolite is also a urinary metabolite of the organophosphorus insecticides parathion and parathion-methyl and has been proposed as a useful index of exposure to these chemicals (Denga et al., 1995); however, it is not likely that there would be simultaneous occupational exposures to nitrobenzene and parathion or parathion-methyl.

Colorimetric methods have been used to determine nitrobenzene metabolites. However, Harmer et al. (1989) noted that colorimetric methods were insensitive, tedious and lacking in specificity. The method for p-nitrophenol and p-aminophenol in urine samples involves acid hydrolysis of conjugates followed by extraction into an organic solvent at pH 4; after re-extraction into dilute ammonium hydroxide, p-nitrophenol is reduced to p-aminophenol by treatment with zinc and hydrochloric acid. Aminophenol is then converted by reaction with phenol to indophenol and the absorbance determined spectrophotometrically. The procedure must be followed exactly, since modification can affect the results (Salmowa et al., 1963; Ikeda & Kita, 1964; Piotrowski, 1967).

Robinson et al. (1951) used a combination of steam distillation, chemical reactions, selective extractions, paper chromatography, silica column chromatography and colorimetric reactions to semiquantitatively estimate *o*-, *m*- and *p*-nitrophenols, nitrobenzene, aniline, 4-nitrocatechol and *o*-, *m*- and *p*-aminophenols in rabbit urine. Similar methods were used by Parke (1956) to estimate the metabolism of [U-^{14}C]nitrobenzene in rabbit and guinea-pig urine.

2)   Other urinary metabolites

Yoshioka et al. (1989) used reversed-phase HPLC with spectrophotometric detection to follow the metabolite hydroxylaminobenzene.

3)   Nitrobenzene in blood

Lewalter & Ellrich (1991) reported a capillary GC method for nitroaromatic compounds, including nitrobenzene, in plasma samples. Nitroaromatic compounds are extracted and simultaneously concentrated using 2,2,4-trimethylpentane (iso-octane), with reported recoveries of 78–119%. GC is carried out in a fused silica capillary with a chemically bonded stationary phase and detected by an electron capture detector. The detection limit was reported as 10 μg/litre of blood.

4)   Methaemoglobin

Exposure to nitrobenzene can lead to the formation of methaemoglobin, and this has been proposed as a biomarker of exposure (ACGIH, 2000). However, this condition in itself cannot be used as a specific marker of exposure, since other toxic substances can have the same effect. For accurate results, blood samples must be treated with an anticoagulant (heparin, ethylenediaminetetraacetic acid [EDTA] or acid–citrate–dextrose solution) and rapidly transported (<1 h) to the assay laboratory. The method is based on the change in absorbance at 630–635 nm after addition of cyanide to convert methaemoglobin to cyanomethaemoglobin (Fairbanks & Klee, 1986).

5)   Haemoglobin adducts

Haemoglobin adducts have been proposed as a marker of exposure to nitrobenzene (Neumann, 1984; Albrecht & Neumann, 1985), and the Deutsche Forschungsgemeinschaft has proposed a 'biological tolerance value for occupational exposures' based on aniline released from haemoglobin adducts (DFG, 1995).

6)   Nitrobenzene in breath

Nitrobenzene and aniline in expired air from rabbits have been determined spectrophotometrically (Parke, 1956).

3. SOURCES OF HUMAN AND ENVIRONMENTAL EXPOSURE

3.1 Natural occurrence

No information on any natural occurrence of nitrobenzene was found.

3.2 Anthropogenic sources

3.2.1 Production

World production of nitrobenzene in 1994 was estimated at 2 133 800 tonnes; about one-third was produced in the USA (Camara et al., 1997).

In the USA, there has been a gradual increase in nitrobenzene production, with the following production/demand amounts, in thousands of tonnes, reported: 73 (1960), 249 (1970), 277 (1980), 435 (1986), 533 (1990) and 740 (1994) (Adkins, 1996; IARC, 1996). Based on increased production capacity and increased production of aniline (the major end-product of nitrobenzene), it is likely that nitrobenzene production volume will continue to increase.

Production of nitrobenzene in Japan was thought to be around 70 000 tonnes in 1980 (Yoshida et al., 1988) and 135 000 tonnes in 1990 (ECDIN, 2000). Patil & Shinde (1989) reported that production of nitrobenzene in India was around 22 000 tonnes per year.

Nitrobenzene is produced at two sites in the United Kingdom with a total capacity of 167 000 tonnes per year. It has been estimated that a maximum of 115 400 tonnes of aniline was produced in the United Kingdom in 1990 (Nielsen et al., 1993). If it is assumed that 98% of the nitrobenzene in the United Kingdom is used to make aniline (as is the case in the USA; see section 3.3), then the total amount of nitrobenzene used in the United Kingdom would be around 155 600 tonnes per year.

Capacities for nitrobenzene production are available for several Western European countries (SRI, 1985) and are shown in Table 3. Production for Western Europe was reported as 670 000 tonnes in 1990 (ECDIN, 2000).

Table 3. Nitrobenzene production capacities in European countries in 1985[a,b]

Country	Capacity (tonnes)
Belgium	200 000
Germany	240 000
Italy	18 000
Portugal	70 000
Switzerland	5 000
United Kingdom	145 000
USA	434 000
Japan	97 000

[a] From SRI (1985).
[b] Production in USA and Japan is described in more detail in the text.

3.2.2 Production processes

Nitrobenzene is produced commercially by the exothermic nitration of benzene with fuming nitric acid in the presence of a sulfuric acid catalyst. The crude nitrobenzene is passed through washer-separators to remove residual acid and is then distilled to remove benzene, water, dinitrobenzene and trinitrobenzene. The reaction can be carried out as either a batch or a continuous process. In the batch process, benzene is nitrated using a nitrating mixture consisting of 53–60% sulfuric acid, 32–39% nitric acid and 8% water. The temperature is maintained at 50–65 °C and then raised to 90 °C towards the end of the reaction (reaction time approximately 2–4 h). The reaction mixture is then run into a separator, where the spent acid settles out and is drawn off and usually recycled. The crude nitrobenzene is drawn off the top of the separator and is then used directly in the manufacture of aniline. If pure nitrobenzene is required, the crude product is washed with water and dilute sodium carbonate and then distilled (Dorigan & Hushon, 1976; Liepins et al., 1977).

In the continuous process, the sequence of operations is the same as in the batch process, but lower nitric acid concentrations and shorter reaction times can be utilized. It is also possible to carry out the reaction as a gas-phase process (Dorigan & Hushon, 1976; Liepins et al., 1977).

3.2.3 *Production losses to the environment*

Most (97–98%) of the nitrobenzene produced is retained in closed systems for use in synthesizing aniline and substituted nitrobenzenes and anilines (Dorigan & Hushon, 1976; Anonymous, 1987). Yoshida et al. (1988) estimated a loss of 0.1% in the production of nitro-benzene. The loss was thought to be mainly to the water phase. Losses to wastewater have been observed to be 0.09% of production in one plant and 2.0% in another (Dorigan & Hushon, 1976). Patil & Shinde (1989) reported that nitrobenzene was present in the wastewater streams from the washers at an Indian nitrobenzene/aniline production plant. The wastewater from the aniline plant was found to contain 1980 mg nitrobenzene/litre, which corresponds to the limit of water solubility.

Pope et al. (1988) gave the following emission factors for losses during the production of nitrobenzene/aniline:

general emission: 8×10^{-6} kg nitrobenzene/kg produced
fugitive emission: 3.1×10^{-4} kg nitrobenzene/kg produced
storage emission: 6.0×10^{-3} kg nitrobenzene/kg produced

Applying these factors to the estimated amount of aniline pro-duced in the United Kingdom in 1990 (see section 3.2.1) gives a maximum emission of 730 tonnes per year in the United Kingdom from this source. Similarly, using the 1990 figure for US nitrobenzene production would give a calculated maximum emission of near 3370 tonnes. Elsewhere, it has been estimated that around 760 tonnes of nitrobenzene are released each year in the USA from the production of nitrobenzene/aniline (White, 1980). Guicherit & Schulting (1985) estimated that around 230 tonnes of nitrobenzene were emitted per year in the Netherlands from stationary sources.

The rate of emission of nitrobenzene to the atmosphere from an industrial wastewater treatment plant has been estimated from flux chamber measurements as 0.100–0.132 mg/min from each square metre of the plant (Gholson et al., 1991).

Emissions of nitrobenzene to the environment are also likely to occur during the manufacture of other chemicals from nitrobenzene. Pope et al. (1988) quoted an emission factor of 1.5×10^{-3} kg nitrobenzene/kg nitrobenzene used for the manufacture of dichloroaniline and dinitrobenzene. The same reference also gave the following general emission factors for use of nitrobenzene as a chemical intermediate:

general emission: 1.05×10^{-3} kg nitrobenzene/kg used
fugitive emission: 3.0×10^{-4} kg nitrobenzene/kg used
storage emission: 1.5×10^{-4} kg nitrobenzene/kg used

It is not known how much nitrobenzene is used to produce chemicals other than aniline. In the United Kingdom, the maximum amount of nitrobenzene used to produce chemicals other than aniline is likely to be around 3100 tonnes per year (Crookes et al., 1994). This would lead to a nitrobenzene emission of around 5 tonnes per year, which is negligible when compared with the emission from production of aniline. It is likely that a similar situation would pertain to other producer countries.

If a significant amount of nitrobenzene is used as a solvent, for instance, for cellulose ethers, then this could represent an important source of nitrobenzene in the environment, as solvent emissions are likely to be higher than production emissions. Around 1.5% of nitrobenzene produced in the USA is used as a solvent for the manufacture of cellulose ethers (see section 3.3).

3.2.4   Non-industrial sources

Nitrobenzene has been shown to be emitted from a multiple-hearth sewage sludge incineration unit in the USA. The unit consisted of 12 hearths and operated at a rate of 13–15 tonnes per hour, with a maximum temperature of 770 °C at the sixth hearth. Nitrobenzene was monitored at the scrubber inlet and outlet. The concentrations measured were 60 μg/m^3 at the scrubber inlet (corresponding to an emission of 3.2 g/h) and 16 μg/m^3 at the scrubber outlet (corresponding to an emission of 0.9 g/h). The scrubber reduced the nitrobenzene concentration by 71% (Gerstle & Carvitti, 1987; Gerstle, 1988).

The levels of nitrobenzene in air have been measured at five hazardous waste landfills and one sanitary landfill in New Jersey, USA. Samples were collected over a 24-h period at five locations within each landfill. Mean levels measured in the five hazardous waste landfills were 0.05, 0.65, 2.7, 1.0 and 6.6 $\mu g/m^3$. The maximum level recorded was 51.8 $\mu g/m^3$. At the sanitary landfill, nitrobenzene was below the detection limit (0.25 $\mu g/m^3$) at all locations (Gianti et al., 1984; Harkov et al., 1985). Similar results have been reported by LaRegina et al. (1986) for the same sites.

Nitrobenzene has been shown to be formed from the atmospheric reactions of benzene in the presence of nitrogen oxides. The reaction is thought to be initiated by hydroxyl radicals (Hoshino et al., 1978; Kenley et al., 1981; Bandow et al., 1985; Spicer et al., 1985; Atkinson, 1990). Nitrobenzene, once formed, reacts quite slowly in the atmosphere (see section 4.2); this could therefore provide a major source of atmospheric nitrobenzene, although it has not been possible to quantify this source. Atkinson et al. (1987) reported that aniline is slowly oxidized to nitrobenzene by ozone. These reactions are summarized in Figure 1.

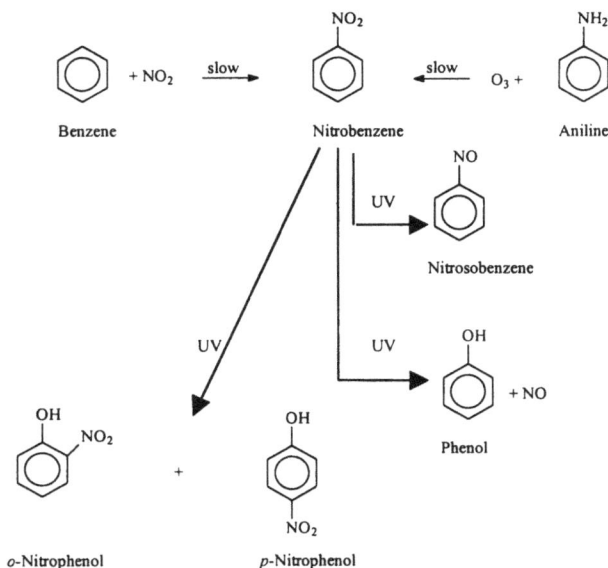

Fig. 1. Atmospheric reactions generating and removing nitrobenzene (ATSDR, 1990).

3.3   Uses

Nitrobenzene is used primarily in the production of aniline, but it is also used as a solvent and as an ingredient in metal polishes and soaps. In the USA, around 98% of nitrobenzene produced is converted into aniline; the major use of aniline is in the manufacture of polyurethanes. Nitrobenzene is also used as a solvent in petroleum refining, as a solvent in the manufacture of cellulose ethers and cellulose acetate (around 1.5%), in Friedel-Crafts reactions to hold the catalyst in solution (it dissolves anhydrous aluminium chloride as a result of the formation of a complex) and in the manufacture of dinitrobenzenes and dichloroanilines (around 0.5%) (Weant & McCormick, 1984; Rogozen et al., 1987). It is also used in the synthesis of other organic compounds, including acetaminophen (ACGIH, 1991).

According to the BUA (1994), nitrobenzene is used in Western Europe for the purposes shown in Table 4.

Table 4. Type and estimated consumption of nitrobenzene in Western Europe in 1979[a]

Main application areas or chemical manufacture	Nitrobenzene consumption (tonnes/year) in Western Europe
Aniline	380 000
m-Nitrobenzenesulfonic acid	5 000
m-Chloronitrobenzene	4 300
Hydrazobenzene	1 000
Dinitrobenzene	4 000
Others (solvents, dyes)	4 000
Total	398 300

[a] From BUA (1994).

Dunlap (1981) reported that most of the production of aniline and other substituted nitrobenzenes from nitrobenzene goes into the manufacture of various plastic monomers and polymers (50%) and rubber chemicals (27%), with a smaller proportion into the synthesis of hydroquinones (5%), dyes and intermediates (6%), drugs (3%), pesticides and other specialty items (9%).

Past minor uses of nitrobenzene included use as a flavouring agent, as a solvent in marking inks and in metal, furniture, floor and shoe polishes, as a perfume, including in perfumed soaps, as a dye intermediate, as a deodorant and disinfectant, in leather dressing, for refining lubricating oils and as a flavouring agent (Polson & Tattersall, 1969; Collins et al., 1982; Rogozen et al., 1987; HSDB, 1988; Hedgecott & Rogers, 1991; Adkins, 1996). It is not known whether it may still be used in some countries as a solvent in some consumer products (e.g., shoe polish).

4. ENVIRONMENTAL TRANSPORT, DISTRIBUTION AND TRANSFORMATION

4.1 Transport and distribution between media

The movement of nitrobenzene in air, water and soil may be predicted by its physical properties (see also chapter 2): water solubility of 1900 mg/litre; vapour pressure of 20 Pa (20 °C); octanol/water partition coefficient (log K_{ow}) of 1.6–2.0; soil/sediment sorption coefficient (K_{sed}) of 36 (Mabey et al., 1982); mean organic matter/water partition coefficient (K_{om}) of 50.1 (Briggs, 1981); and Henry's law constant of 0.9–2.4 Pa·m³/mol.

The following distribution of nitrobenzene in the environment at 20 °C was calculated using Mackay's Level I fugacity model assuming equilibrium conditions: air, 36%; water, 62%; soil and sediment, 2%; and aquatic biomass, 0% (BUA, 1994).

4.1.1 Air

The vapour pressure of nitrobenzene of 20 Pa at 20 °C indicates that nitrobenzene may be slightly volatile. Cupitt (1980) estimated that washout by rainfall (either through solution in raindrops or by removal of nitrobenzene sorbed onto particulates) and dryfall of particulates is negligible, as expressly measured in field releases (Dana et al., 1984). Atmospheric residence time was estimated to be 190 days (Cupitt, 1980).

Vapour densities reported for nitrobenzene relative to air range from 4.1 to 4.25 (Dorigan & Hushon, 1976; Beard & Noe, 1981; G.E. Anderson, 1983; Adkins, 1996). Removal processes for nitrobenzene in air may involve settling of vapour due to its higher density relative to air (Dorigan & Hushon, 1976).

4.1.2 Water

The Henry's law constant values (0.9–2.4 Pa·m³/mol) suggest that nitrobenzene volatilizes slowly from the water phase (BUA, 1994);

transfer from water to air will be significant, but not rapid (Lyman et al., 1982).

The half-life for evaporation of nitrobenzene has been estimated as 8–20 days from two major lakes in Istanbul, Turkey (Ince, 1992). Zoeteman et al. (1980) estimated the half-life for volatilization of nitrobenzene from river water to be 1 day.

Piwoni et al. (1986) found that nitrobenzene did not volatilize in their microcosms simulating land application of wastewater, but was totally degraded. In contrast, in a laboratory-scale waste treatment study, Davis et al. (1981) estimated that 25% of the nitrobenzene was degraded and 75% was lost through volatilization in a system yielding a loss of about 80% of initial nitrobenzene in 6 days. In a stabilization pond study, the half-life by volatilization was about 20 h, with approximately 3% adsorbed to sediments (Davis et al., 1983). The EXAMS computer model (Burns et al., 1982) predicts volatilization half-lives of 12 days (river) to 68 days (eutrophic lake) and up to 2% sediment sorption for nitrobenzene.

4.1.3   Soil and sediment

The sorption behaviour of nitrobenzene has been investigated in several experimental studies using a variety of different soil types. Overall, only a moderate adsorption potential was observed, the amount of which was particularly governed by the soil organic carbon content. Thus, the substance is expected to be relatively mobile in most kinds of soil.

Løkke (1984) studied the adsorption of nitrobenzene onto two soils with organic carbon contents of 1.82% and 2.58%. After shaking the soil slurries with nitrobenzene (applied concentrations 2–100 mg/litre) for up to 72 h, the soil organic carbon/water partition coefficients (K_{oc}) according to the non-linear Freundlich's equation were in the range of 170–370. The amount of nitrobenzene adsorbed was largely determined by the organic carbon content of the soils, whereas the pH, cation exchange capacity and incubation temperature (5 °C and 21 °C) were of minor importance.

Soil sorption coefficients (K_{om} and K_{oc}) of nitrobenzene were investigated by Briggs (1981) using soil samples from four fields (silty loam). The soils had organic matter contents of between 1.09% and 4.25%, and pH values ranged from 6.1 to 7.5. Nitrobenzene, added at concentrations of 5, 10, 15 and 20 mg/litre, was shaken with a soil/ water mixture for 2 h at 20 °C. The determined adsorption isotherms were linear over the applied concentration range. The mean K_{om} was 50, and the mean K_{oc} was 86.

Using three soil columns containing soils with organic carbon contents of 0.2%, 2.2% or 3.7%, K_{oc} values of 30.6, 88.8 and 103 were determined for nitrobenzene (Seip et al., 1986). K_{oc} values of 86.4 (Jeng et al., 1992) and 132 (Koch & Nagel, 1988) have also been quoted for nitrobenzene.

An average K_{oc} value of 96.8 has been determined for nitrobenzene using both a silt loam soil (1.49% organic carbon) and a sandy loam (0.66% organic carbon). The soils were shaken with an aqueous mixture of 16 chemicals, including nitrobenzene (concentration equivalent to 100–650 µg/g soil), for 18 h, and the K_{oc} values obtained for nitrobenzene were 89.0 for the silt loam and 105.6 for the sandy loam (Walton et al., 1989, 1992).

Haderlein & Schwarzenbach (1993) studied the adsorption of substituted nitrobenzenes to various mineral sorbents. Nitrobenzene was found to adsorb specifically to the negatively charged siloxane sites. The strength of sorption was dependent on the type of cation present in the mineral, the adsorption being stronger when the more weakly hydrated cations (e.g., NH_4^+, K^+, Rb^+ or Cs^+) were present.

Roy & Griffin (1985) calculated K_{oc} values for nitrobenzene of 79 (from solubility) and 62 (from K_{ow}). Using these values, the authors classified nitrobenzene as highly mobile in soils. A K_{oc} value of 148 has been calculated for nitrobenzene using molecular connectivity indices (Bahnick & Doucette, 1988).

Briggs (1981) compared the soil sorption coefficient (K_{sed}), expressed in terms of organic matter (K_{om}), where $K_{om} = 100 \times K_{sed}/(\%$ organic matter), with the K_{ow} for a wide variety of chemicals and soils. Briggs (1973) classified soil mobility using log K_{ow} and log organic matter content and compared this classification with that of Helling & Turner (1968), based on soil thin-layer chromatography. This would suggest that nitrobenzene would be in the mobility class III (intermediate).

Jury et al. (1984) also classified nitrobenzene as intermediately mobile, but noted that its loss from soil would be enhanced by evaporation of water. Moreover, because nitrobenzene has relatively poor diffusive flux, the material would tend to move as a bolus within soil. It was hypothesized that a deposit 10 cm deep in soil would have a volatilization half-life of about 19 days.

Other results also indicate that nitrobenzene is intermediately mobile in forest and agricultural soils (Seip et al., 1986). However, nitrobenzene was somewhat more mobile in soil with lower organic carbon content.

No data were found on adsorption of nitrobenzene to sediment.

4.2 Abiotic degradation

4.2.1 Air

4.2.1.1 Direct photolysis

The direct photolysis pathways of nitrobenzene in air are shown in Figure 1 in chapter 3 (according to US EPA, 1985). *p*-Nitrophenol and nitrosobenzene were reported to be the principal photodegradation products of nitrobenzene vapours exposed to ultraviolet (UV) light in air (Hastings & Matsen, 1948). In another study, both *o*- and *p*-nitrophenols were found when oxygen was present, and phenol was found when oxygen was absent (Nojima & Kanno, 1977).

Under laboratory conditions, direct photolysis of nitrobenzene in solvents such as isopropanol yields hydroxylaminobenzene, which can be oxidized to nitrosobenzene by oxygen (Hurley & Testa, 1966, 1967). Hydroxylaminobenzene and nitrosobenzene can then combine to form azoxybenzene. However, these reactions may not be important under natural conditions in the absence of hydrogen donors (Mabey et al., 1982).

With light of wavelength below 290 nm (UV), Hendry & Kenley (1979) measured the direct photolysis rate of nitrobenzene at 30°N latitude as 7.6 per day at the summer solstice and 0.7 per day at the winter solstice. These rates correspond to lifetimes of <1 day, except near to the winter solstice. It was estimated by the authors that the process should be possible up to wavelengths of approximately 304 nm, which is just in the solar spectrum.

After irradiation of ^{14}C-labelled nitrobenzene adsorbed onto silica gel (60 ng nitrobenzene/g adsorbent) with a UV lamp (wavelength ~290 nm) for 17 h, 6.7% of the applied radioactivity was detected as carbon dioxide (Freitag et al., 1982, 1985).

4.2.1.2   Indirect photolysis (photo-oxidation)

The most important reactants concerning photo-oxidation in air are hydroxyl radicals and ozone. Nitrobenzene has been shown to react slowly with hydroxyl radicals under simulated atmospheric conditions. Measured reaction rate constants are shown in Table 5. Rate constants for the reaction of nitrobenzene with ozone have also been determined and are shown in Table 6. This reaction is significantly slower than the one with hydroxyl radicals (often slower than the decomposition of ozone under the experimental conditions used), and so it is only possible to estimate an upper limit for the rate constant.

Table 5. Rate constants for reaction of nitrobenzene with hydroxyl radicals

Reaction rate constant (cm^3/molecule per second)	Reference
$<7 \times 10^{-13}$	Atkinson et al., 1987
2.1×10^{-13}	Atkinson, 1985
1.4×10^{-13}	Witte et al., 1986
6×10^{-14a}	Cupitt, 1980
9.1×10^{-14}	Witte & Zetzsch, 1984
$<4 \times 10^{-13}$	Arnts et al., 1987

[a] Calculated value.

Table 6. Rate constants for reaction of nitrobenzene with ozone

Reaction rate constant (cm^3/molecule per second)	Reference
$<1 \times 10^{-20}$	Atkinson, 1990
$<7 \times 10^{-21}$	Atkinson et al., 1987
$<5 \times 10^{-23a}$	Cupitt, 1980

[a] Calculated value.

Assuming that these are first-order reactions, and based on an average concentration in the lower troposphere of about 6×10^5 hydroxyl radicals/cm^3 (BUA, 1995) and 1.25 ozone molecules/cm^3 (0.1 mg/m^3; BUA, 1993), the half-lives ($t_{1/2}$) can be calculated as follows:

$t_{1/2}$ (OH) = 19–223 days
$t_{1/2}$ (O$_3$) = 1.8–352 years

These calculations are roughly similar to the estimations of Atkinson et al. (1987), who projected half-lives of nitrobenzene of 180 days by reaction with hydroxyl radicals and more than 6 years by reaction with ozone in "clean" air. In typical, moderately "dirty" air, these values would decrease to 90 days and more than 2 years, respectively.

Photochemical oxidation of nitrobenzene by hydrogen peroxide yields *p*-, *o*- and *m*-nitrophenols (Draper & Crosby, 1984), with an estimated half-life of 250 days (Dorfman & Adams, 1973).

Spicer et al. (1985) studied the reaction of nitrobenzene in a smog chamber containing a propylene/butane/nitrogen dioxide mixture at 30 °C. The lifetime of nitrobenzene under the conditions of the experiment was around 50 h, and this was thought to represent a lifetime of around 4–5 days in the atmosphere. The major products formed were *o*- and *p*-nitrophenol, plus smaller amounts of other phenolic compounds.

The long lifetime means that nitrobenzene is likely to be transported large distances in the atmosphere from the source of release. As it does not contain halogen atoms, nitrobenzene will have a low potential for ozone depletion.

No information has been found on the global warming potential of nitrobenzene.

4.2.2   *Water*

4.2.2.1   *Hydrolysis*

Measured data on the hydrolytic degradation of nitrobenzene are not available. From its structure and its chemical properties, the substance should not decompose in water (BUA, 1994).

However, nitrobenzene has been shown to be reduced by hydrogen sulfide in aqueous solution. The reaction was found to be mediated by natural organic matter (quinone-type moieties may be important) from a variety of sources (Dunnivant et al., 1992). Similarly, Schwarzenbach et al. (1990) found that nitrobenzene was reduced to aniline in solution by hydrogen sulfide or cysteine in the presence of naturally occurring quinones or water-soluble iron porphyrin. Both the quinones and iron porphyrin were thought to act as electron carriers during the reaction. Reactions of this type have also been noted with sulfide minerals (see section 4.2.3).

4.2.2.2   *Direct photolysis*

By direct photolysis, nitrobenzene has a half-life of 2.5 to more than 6 days near the surface of bodies of water in the vicinity of 40°N latitude (Zepp & Schlotzhauer, 1983).

4.2.2.3   *Indirect photolysis (photo-oxidation)*

The rate constants for reaction of nitrobenzene with various radical species in water at room temperature are compiled in Table 7.

Typical values for the concentration of hydroxyl radicals in surface waters range between a low of 5.0×10^{-19} mol/litre to a high of 2.0×10^{-17} mol/litre (Howard et al., 1990). Assuming first-order kinetics, this means that typical half-lives estimated for nitrobenzene in water, based solely on reaction with hydroxyl radicals, would be between 125 days and 13 years at pH 7. No information was found on the concentrations of hydrated electrons and hydrogen atoms in natural waters, but it is expected that their concentrations would be much lower than that of hydroxyl radicals. Therefore, it is likely that reaction with all three radical species will be only a minor removal pathway for nitrobenzene in water. However, Zepp et al. (1987a) reported that hydrated electrons from dissolved organic matter could significantly increase photoreduction of compounds such as nitrobenzene and also that photolysis of nitrate ions to hydroxyl radicals increased nitrobenzene photodegradation (Zepp et al., 1987b).

Table 7. Rate constants for reaction of nitrobenzene
with various species in water

Reactive species	pH	Rate constant (litres/mol per second)	Reference
Hydrated electrons	7	3.0×10^{10}	Anbar & Neta, 1967
Hydrogen atoms	7	1.7×10^{9}	Anbar & Neta, 1967
Hydroxyl radicals	1	4.7×10^{9}	Dorfman & Adams, 1973
	7	3.2×10^{9}	Dorfman & Adams, 1973
	9	2.0×10^{9}	Anbar & Neta, 1967
	9	3.4×10^{9}	Dorfman & Adams, 1973
	10.5	1.5×10^{9}	Anbar & Neta, 1967
	10.5	2.2×10^{9}	Dorfman & Adams, 1973

Callahan et al. (1979) proposed that sorption of nitrobenzene to humic acids could enhance the photolytic destruction of nitrobenzene. However, the rate of photolysis of nitrobenzene in solution using natural sunlight or monochromatic light (wavelength of 313 or 366 nm) was not significantly affected by the presence of humic acids, and a near-surface half-life of 133 days was estimated for nitrobenzene photolysis at 40°N latitude (Simmons & Zepp, 1986). The presence of algae (several species) also did not enhance the photolysis of nitrobenzene in solutions exposed to sunlight for 3–4 h (Zepp & Schlotzhauer, 1983).

4.2.3   Soil and sediment

A study of the efficacy of soil infiltration along the river Rhine in the Netherlands showed that nitrobenzene was removed completely when passed continuously through 50 cm of a peat sand artificial dune in infiltration basins (Piet et al., 1981).

Wolfe (1992) reported that nitrobenzene was reduced to aniline by abiotic mechanisms in a variety of sediment samples collected from local ponds and streams. A half-life for the reaction of 56 min was found for a sediment:water ratio of 0.13.

For both studies, it is not possible to identify or even quantify the underlying mechanisms (biotic, abiotic) that were responsible for the removal or transformation of the applied nitrobenzene.

Nitrobenzene has been shown to be reduced by several sulfide minerals (at a mineral concentration of 0.24 mol/litre) under aerobic conditions. The half-lives for the disappearance of nitrobenzene were found to be 7.5 h for reaction with sodium sulfide, 40 h with alabandite (manganese sulfide), 105 h with sphalerite (zinc sulfide) and 360 h with molybdenite (molybdenum sulfide). Aniline was identified as the major reaction product with sodium sulfide, molybdenite and alabandite. Several unidentified products were formed in the reaction with sphalerite. The mineral solubility and dissolution rate were found to be the key factors in determining the rate of reaction (Yu & Bailey, 1992).

4.3   Bioaccumulation

4.3.1   Aquatic species

The fate of nitrobenzene has been studied in a simple model "farm pond" aquatic ecosystem with a six-element food-chain. The system contained phytoplankton and zooplankton, green filamentous algae (*Oedogonium cardiacum*), snails (*Physa* sp.), water fleas (*Daphnia magna*), mosquito larvae (*Culex quinquifasciatus*) (fourth instar) and mosquito fish (*Gambusia affinis*). At the start of the experiment, 300 daphnia, 200 mosquito larvae, 6 snails, strands of algae and miscellaneous plankton were exposed to ^{14}C-labelled nitrobenzene at a concentration of 0.01–0.1 mg/litre. After 24 h, 50 mosquito larvae and 100 daphnia were removed and 3 fish were added. The experiment was terminated after a further 24 h. The ecological magnification index (ratio of the concentration of parent material in the organism to the concentration of parent material in water) was about 8 in mosquito fish after a 24-h exposure. Bioaccumulation from water is not considered significant at these values (Trabalka & Garten, 1982). The ecological magnification index was 0.7 in snails, 0.8 in mosquito larvae, 0.15 in water fleas and 0.03 in green algae. Thus, nitrobenzene remained mainly in the water during the experiment and was neither stored nor ecologically magnified. It was also found to be reduced to aniline in all organisms and acetylated in fish, whereas the mosquito larvae and snails also hydroxylated it to nitrophenols; however, the extent of metabolism was small (Lu & Metcalf, 1975). It should be noted that the ecological magnification index may not be equivalent to a bioconcentration factor (BCF), which usually assumes that equilibrium between water and the organism has been reached.

The bioaccumulation of ^{14}C-labelled nitrobenzene has been studied in algae (*Chlorella fusca*) and fish (golden ide [*Leuciscus idus melanotus*]). The nitrobenzene concentration used was 50 µg/litre, and exposure occurred for 1 day (algae) or 3 days (fish). The BCFs measured on a wet weight basis were 24 for the algae and <10 for the fish (Geyer et al., 1981, 1984; Freitag et al., 1982, 1985).

Guppies (*Poecilia reticulata*) were exposed to nitrobenzene (12 mg/litre) for 3 days (by which time equilibrium was reached), with solutions renewed each day. A BCF (fat weight) of 29.5 was determined (Deneer et al., 1987); using the average fat content of the guppies (8%), this value can be converted to a whole body weight BCF of 2.4.

Veith et al. (1979) reported a BCF of 15 in the fathead minnow (*Pimephales promelas*) for a 28-day exposure to nitrobenzene in a flow-through test. A less satisfactory 3-day static measurement gave a BCF of less than 10 for the golden orfe (*Leuciscus idus*) (Freitag et al., 1982). Nitrobenzene has been shown to have low bioconcentration potential in carp (*Cyprinus carpio*) (Kubota, 1979).

In a bioaccumulation study performed according to Organisation for Economic Co-operation and Development (OECD) Guideline 301C, carp (*Cyprinus carpio*) were exposed to nitrobenzene concentrations of 0.125 and 0.0125 mg/litre. After 6 weeks of incubation, BCFs in the range of 1.6–7.7 (0.0125 mg/litre) and 3.1–4.8 (0.125 mg/litre) were determined. No information is given as to whether the BCFs were measured under steady-state conditions (MITI, 1992).

In conclusion, the measured BCFs for nitrobenzene in a number of organisms indicate minimal potential for bioaccumulation.

4.3.2 *Terrestrial plants*

Nitrobenzene may bioconcentrate in terrestrial plants. The relatively rapid uptake of ^{14}C-labelled nitrobenzene into mature soybean

(*Glycine max*) plants was reported by McFarlane et al. (1987a, 1987b) and Nolt (1988). The majority of the nitrobenzene taken up by the plants was associated with the roots, and very little was transported to other areas of the plant. It was concluded that nitrobenzene was probably metabolized within the root system (McCrady et al., 1987; McFarlane et al., 1987a, 1987b).

The roots of soybean (*Glycine max*) plants were exposed to a hydroponic solution containing ^{14}C-labelled nitrobenzene at 0.02–100 µg/ml for a 72-h exposure period. The plants were dissected into roots and shoots and analysed for ^{14}C label and for nitrobenzene. At the highest concentration, radioactivity was almost evenly distributed between the roots and shoots; at the lower doses, on the other hand, approximately 80% of the radioactivity was in the roots (Fletcher et al., 1990). It should be noted that this experiment was a water-only exposure; consequently, nitrobenzene is likely to be more bioavailable to plants than when exposures are via water and soil.

4.4 Biotransformation

The biodegradation of nitrobenzene has been studied using a wide variety of mixed microbial consortia as well as single bacterial species from different environmental compartments. In general, either non-acclimated inocula were unable to use nitrobenzene as a sole source of carbon and energy or degradation occurred only after extended acclimation periods. Acclimated microorganisms, particularly from industrial wastewater treatment plants, however, showed elimination of up to 100% within incubation periods of 1–5 days.

4.4.1 Aerobic degradation

4.4.1.1 Biodegradation by non-acclimated microorganisms

In a modified MITI-Test (I) conducted according to OECD Guideline 301C (OECD, 1981), the mineralization of nitrobenzene by a mixed microbial inoculum sampled from different sewage plants, rivers and a bay was determined by measurement of the biochemical oxygen demand (BOD). Incubation of an initial test substance concentration of 100 mg/litre with a sludge concentration of 30 mg/litre (non-adapted) at 25 °C and pH 7.0 resulted in 3.3% degradation after

14 days (MITI, 1992). In another MITI ready biodegradability test, Kubota (1979) found no degradation of the applied nitrobenzene.

Alexander & Lustigman (1966) studied the primary degradation of nitrobenzene as the sole source of carbon by a natural microbial population of an uncontaminated silty loam soil (Niagara, USA). Soil suspensions with initial nitrobenzene concentrations in the range of 5–10 mg/litre and inoculated with a low inoculum density (1 ml of a 1% loam suspension) were incubated in the dark at 25 °C. Primary degradation (ring cleavage) was monitored by measuring the UV absorption in the wavelength range of 250–268 nm. No significant ring cleavage was detected in the batches for nitrobenzene after 64 days of incubation.

Incubation of nitrobenzene (50 µg/litre) with activated sludge from a municipal sewage treatment plant (1 g dry weight/litre) was carried out for 5 days at 25 °C. Meat extract and peptone were added as additional substrates, and mineralization of nitrobenzene was monitored by carbon dioxide analysis. No metabolites were identified during the experiment, and only 0.4% of the applied radioactivity was found as carbon dioxide, indicating that nitrobenzene was not metabolized to any significant extent under the conditions of the test (Freitag et al., 1982, 1985).

Several other authors have reported that nitrobenzene was not significantly degraded by non-acclimated microorganisms under various conditions. Nitrobenzene was reported either to be highly resistant to degradation or to inhibit biodegradation of other components of the waste in several biodegradation studies (Marion & Malaney, 1963; Lutin et al., 1965; Barth & Bunch, 1979; Davis et al., 1981; Korte & Klein, 1982). However, these effects were observed at concentrations of nitrobenzene greater than or equal to 50 mg/litre, much higher than those detected in ambient waters.

Urano & Kato (1986) found that nitrobenzene (100 mg/litre) was not significantly degraded within 14 days of incubation (20 °C; mineral salts medium) with a non-acclimated activated sludge inoculum (sludge concentration 30 mg/litre).

4.4.1.2   *Biodegradation by acclimated microorganisms*

Pseudomonas pseudoalcaligenes, which is able to use nitroben-zene as the sole source of carbon, nitrogen and energy, was isolated from soil and groundwater contaminated with nitrobenzene. The range of aromatic substrates able to support growth was limited to nitro-benzene, hydroxylaminobenzene and 2-aminophenol. Nitrobenzene, nitrosobenzene, hydroxylaminobenzene and 2-aminophenol stimulated oxygen uptake in resting cells and in extracts of nitrobenzene-grown cells. Washed suspensions of nitrobenzene-grown cells removed nitrobenzene from culture fluids with the concomitant release of ammonia. It was found that nitrobenzene was reduced to hydroxyl-aminobenzene via nitrosobenzene. Under aerobic and anaerobic condi-tions, the hydroxylaminobenzene undergoes an enzyme-catalysed reaction to 2-aminophenol, which undergoes a ring cleavage to produce 2-aminomuconic semialdehyde, with the release of ammonia (Nishino & Spain, 1993). A proposed pathway for the biodegradation of nitrobenzene is given in Figure 2.

Cultures of several species of *Pseudomonas* were grown at 30 °C with nitrobenzene supplied in the vapour phase above the culture. The bacteria included *P. putida*, *P. pickettii*, *P. cepacia*, *P. mendocina* and several unidentified strains. All the bacteria were known to contain toluene degradative pathways. The cells were harvested and then incubated with nitrobenzene at 30 °C to enable metabolites to be iden-tified. All the strains grew in the presence of nitrobenzene vapour when glucose or arginine were provided as an alternative carbon source, but none grew on nitrobenzene as the sole carbon source. Several metabolites were identified from the various strains, including 3-nitrocatechol, 4-nitrocatechol, *m*-nitrophenol and *p*-nitrophenol, although several strains did not transform nitrobenzene at all. The nitrocatechols were slowly degraded to unidentified metabolites. Results indicate that the nitrobenzene ring is subject to initial attack by both mono- and dioxygenase enzymes (Haigler & Spain, 1991). In contrast to this, Smith & Rosazza (1974) found no phenolic metabolites when nitrobenzene (1000 mg/litre) was incubated at 27 °C for 24–72 h with the following microorganisms with demonstrated aromatic metabolizing ability: *Aspergillus niger*, *Penicillium chrysogenum*, *Cunninghamella blakesleeana*, *Aspergillus ochraceous*, *Gliocladium deliquescens*, *Streptomyces* sp., *Rhizopus stolonifer*, *lunata*, *Streptomyces rimosus*, *Cunninghamella bainieri* and *Helicostylum piriforme*.

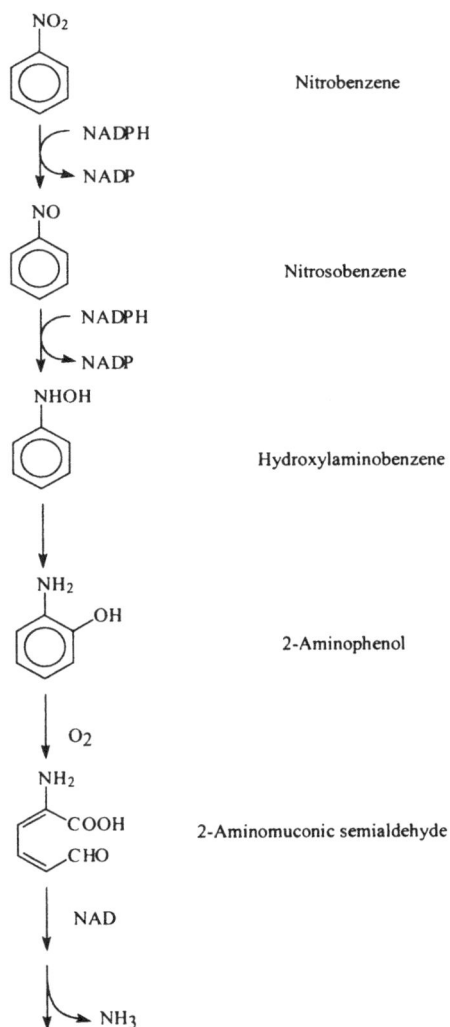

Fig. 2. Proposed pathway for the biodegradation of nitrobenzene in *Pseudomonas pseudoalcaligenes* (Nishino & Spain, 1993)

Gomółka & Gomółka (1979) investigated the ability of micro-organisms in municipal wastewater to synthesize enzymes for the catalytic transformation of different nitrobenzene concentrations. Three concentration ranges (5–100, 50–300 and 400–1400 mg/litre) were aerated in a respirometer, with continual analysis of oxygen consumption, pH and nitrobenzene content. The acclimation time — i.e., the time for 30–40% degradation in terms of oxygen consumption — was in the range of 2–5 days for initial nitrobenzene concentrations of 50–300 mg/litre. After the acclimation period, concentrations of nitrobenzene up to 300 mg/litre were degraded slowly. No adverse effect on microbial respiration was observed at nitrobenzene concentrations of up to 100 mg/litre, but complete inhibition of oxygen consumption was found at concentrations above 1000 mg/litre. Nitrobenzene was also shown to be degraded using municipal sludge reactors (Gomółka & Gomółka, 1979; Gomółka & Gomółka, 1983).

Davis et al. (1981) investigated the degradation of nitrobenzene and metabolite formation during the decomposition of nitrobenzene using seed from industrial wastewater treatment units and from municipal activated sludge. The industrial sludge contained mainly the four bacterial genera *Acinetobacter*, *Alcaligenes*, *Flavobacterium* and *Pseudomonas* and the yeast *Rhodotorula*. The municipal sludge was not classified for microbial genera. In each experiment, the bacterial cell concentration was 18×10^8 cells/ml, and all incubations were carried out at 28 °C. Degradation was monitored by oxygen uptake measurement in a Warburg respirometer and by substance-specific analysis (GC/MS). Respiration was inhibited by nitrobenzene concentrations of 100 and 200 mg/litre (industrial sludge) and 200 mg/litre (municipal sludge). Using the municipal seed, an initial concentration of nitrobenzene of 50 mg/litre was reduced to 0.3 mg/litre within 6 days. After subtracting the volatile fraction, approximately 20% of removal could be attributed to microbial degradation. After 6 days, a further 50 mg nitrobenzene/litre was added to the flask, and 40 mg/litre was found to remain after a further 6 days' incubation. This was thought to indicate that microbial degradation was occurring mainly via co-metabolism, as the amount of glucose available in the culture was minimal for the second 6-day period. Using the industrial seed (in the endogenous growth phase) and an initial nitrobenzene concentration of 50 mg/litre, approximately 9–10 mg nitrobenzene/litre was biodegraded in 6 days. Aniline and phenol were detected as metabolites. Because of the small decrease in nitrobenzene levels found during the investigations, further experiments were performed to determine whether the applied test substance was adsorbed or absorbed by the bacterial mass. However, after cell digestion, no nitrobenzene could be found in the inocula, indicating that the removal was due to microbial transformation rather than simple adsorption to the culture.

Patil & Shinde (1988) studied the elimination of nitrobenzene both alone and as a mixture with aniline by activated sludge derived from a domestic sewage treatment plant. The inoculum was acclimated for 15 days to wastewater containing both aniline and nitrobenzene. Decomposition was followed by measurement of the chemical oxygen demand (COD) and substance-specific analysis (GC). Initial nitrobenzene concentrations in the range of 184–250 mg/litre were found to be completely degraded in all experiments within 7–8 h of incubation.

Tabak et al. (1981) studied the degradation of nitrobenzene in a static culture procedure using a settled domestic wastewater as microbial inoculum and yeast extract as additional substrate. Initial concentrations of 5 and 10 mg/litre were incubated for 7 days in static culture (25 °C; dark). Subsequently, three subcultures were taken at weekly intervals and incubated under the same conditions as the first culture in order to examine the effects of acclimation. Degradation was monitored by measuring the dissolved organic carbon (DOC) and the total organic carbon (TOC) and by substance-specific analysis (GC). At 5 mg/litre, 100% of the applied nitrobenzene was degraded in all cultures. At 10 mg/litre, 87% was degraded in the original culture, 97% in the first subculture and 100% in the subsequent subcultures.

Under aerobic conditions, nitrobenzene was completely eliminated within about 10 days by samples of raw sewage from a municipal treatment plant. The primary effluent was amended with 10 mg nitrobenzene/litre and was incubated in the dark at 29 °C. Every 7 days, fresh sewage was added to provide additional nutrients. Primary degradation of nitrobenzene was measured by a UV spectrophotometric method. No aromatic amine metabolites were detected (Hallas & Alexander, 1983).

Pitter (1976) performed studies on the biodegradability of nitrobenzene by acclimated activated sludge in a static die-away system equivalent to the Zahn-Wellens test. Prior to the tests, the sludge was gradually acclimated over 20 days to the initial nitrobenzene concentration of 200 mg COD/litre as the sole source of carbon. The test substance was then incubated with the thickened activated sludge (100 mg dry weight/litre) in mineral salts medium at a temperature of 20 ± 3 °C. After 5 days of incubation, 98% of the applied nitrobenzene was degraded (based on COD removal). The specific degradation rate was reported to be 14 mg COD/g dry substance per hour.

Nitrobenzene has been shown to be removed in a wastewater treatment plant that received approximately 50% industrial waste. Nitrobenzene was spiked into the raw wastewater entering the treatment plant at a level of around 0.5 mg/litre. Samples of primary and secondary sludges from this plant were then taken to be used as a feed sludge for a laboratory anaerobic digester, but nitrobenzene was not detected in this feed sludge, indicating that complete removal had occurred in the aerobic treatment plant (Govind et al., 1991).

4.4.1.3   *Degradation of nitrobenzene in soil*

Kincannon & Lin (1985) studied the primary degradation of nitrobenzene in columns containing soil material and waste sludges. Three soil types were used, ranging from clay to sandy soils. Waste sludges (a dissolved air flotation sludge, a slop oil sludge and a wood preserving sludge) were applied to the top of the column and worked into the top 20 cm. The removal of the initial influent concentration of 2400 mg nitrobenzene/kg was monitored by GC analysis in different depths of the soil. Half-lives for nitrobenzene were found to be 56 days in the dissolved air flotation sludge-amended column, 13.4 days in the slop oil sludge-amended column and 196.6 days in the wood preserving sludge-amended column. The contribution of abiotic removal mechanisms remains unclear, as nitrobenzene was found to be removed fairly rapidly from sterilized soil columns, presumably by volatilization, with a half-life of around 9 days.

Anderson et al. (1991) studied the removal of nitrobenzene in two soils, a silt loam of 1.49% organic carbon content and a sandy loam of 0.66% organic carbon content. The experiment was carried out using both sterile and non-sterile soils to distinguish biotic losses from abiotic losses. Nitrobenzene was added to the soils at 100 mg/kg dry weight, and the mixture was incubated at 20 °C in the dark. The half-life of nitrobenzene in both soils was around 9 days, and the differences in the rate of disappearance between sterile and non-sterile soils was slight, indicating that the loss was caused by abiotic processes. Being unable to decisively explain the fate of the applied test substance, the authors discussed irreversible partitioning to soil organic matter and losses during preanalysis storage as possible sinks. Very similar results have also been reported by Walton et al. (1989).

Wilson et al. (1981) studied the fate of nitrobenzene in soil columns packed to a depth of 140 cm with a sandy soil type in a manner that preserved the characteristics of the original profile. The columns received spring water spiked with a mixture of compounds, including nitrobenzene (0.92 and 0.16 mg/litre). Removal of the applied nitrobenzene was measured by substance-specific analysis in the effluent. The feed solutions were applied for 45 days, but a steady-state concentration in the effluent was reached in 25 days. At equilibrium, 60–80% of the nitrobenzene applied to the influent was found in the column effluent, none volatilized, and the remainder (20–40%) was presumed to have degraded on the column. The study indicates the partial degradation of nitrobenzene in soil and confirms its mobility in soil and its potential to leach to groundwater.

The removal of nitrobenzene was determined in a complete-mix, bench-scale, continuous-flow activated sludge reactor fed a synthetic wastewater containing a mixture of readily degradable compounds as well as the compound under study. The activated sludge was sampled from a municipal treatment plant and acclimated to the nitrobenzene-containing wastewater prior to the test. The reactors were operated with a hydraulic retention time of 8 h. Following acclimation, samples were collected over a test period of 60 days and monitored for 5-day biochemical oxygen demand (BOD_5), TOC, COD and nitrobenzene (GC analysis). About 76–98% of the concentration of nitrobenzene applied to the influent (100 mg/litre) was removed during the column passage (Stover & Kincannon, 1983).

Piwoni et al. (1986) used microcosms designed to simulate a rapid-infiltration land treatment system for wastewater to determine the fate of nitrobenzene under such conditions. The microcosms consisted of 1.5-m soil columns with sampling ports at various depths, with the top of the column enclosed in a "greenhouse" through which air was flushed, such that the air was replaced every 8 min. The columns were filled with fine sandy soil, planted to Reed Canary Grass and maintained at 20 ± 2 °C. During incubation, they received water containing a mixture of several chemicals (nitrobenzene concentration 271 µg/litre) each day. Substance-specific analysis (GC) showed that <0.1% of the nitrobenzene volatilized from the column or was found in the final effluent, implying that >99.9% was degraded during passage in the column.

Nitrobenzene (120 mg/litre) was completely degraded within 72 h of incubation with organisms isolated from soil sampled near a chemical manufacturing site (Charde et al., 1990).

In summary, several investigations have reported that nitrobenzene was not readily degraded by activated sludge inoculum. However, concentrations of nitrobenzene used in these studies were generally much higher than those detected in effluents and likely to be toxic to microorganisms. A more extensive range of other studies indicated that the use of raw sewage sludge from wastewater treatment plants can lead to complete degradation of nitrobenzene under aerobic conditions.

4.4.2 Anaerobic degradation

Raw sewage from a municipal treatment plant was amended with 10 mg nitrobenzene/litre and was incubated anaerobically in the dark at 29 °C; every 7 days, fresh sewage was added to provide additional nutrients. Under anaerobic conditions, nitrobenzene was completely removed within 14 days; 50% UV absorption at the nitrobenzene wavelength was observed after 14 days, but GC/MS analysis subsequently showed that this was due not to nitrobenzene but to one of its metabolites. Aniline was detected as the major metabolite (Hallas & Alexander, 1983).

Chou et al. (1978) studied the primary degradation of nitrobenzene by methanogenic bacteria. The bacteria were enriched from a seed of well digested domestic sludge that was grown on acetate for several years. Degradation was tested in reactors for 20 days and in upflow anaerobic filters for 2–10 days. No further details of the test design were reported. The batches were inoculated with methanogenic acetate enriching cultures, and, after substrate utilization had started, inorganic salts, acetate and nitrobenzene were supplied daily. Removal of nitrobenzene was followed by monitoring of the COD or the TOC.

Nitrobenzene at a concentration of 350 mg/litre was found to be degraded after a long acclimation period (81% removal in 110 days).

Degradation of nitrobenzene has been shown to occur in anaerobic, expanded-bed, granular activated carbon reactors. The reactors were used to treat synthetic wastewater containing several semivolatile organic compounds (nitrobenzene at 100 mg/litre). The reactors were gradually acclimated to increasing concentrations of the semivolatile organic compounds over 150 days. The synthetic wastewater was then added at a rate of 8 litres/day (contact time of the reactor was 10.5 h); nitrobenzene was not detected in the reactor effluent. Aniline was detected as a degradation product (Narayanan et al., 1993).

Canton et al. (1985) measured an 8% decrease in nitrobenzene after 8 days in anaerobic culture containing unadapted inoculum, but reported a half-life of less than 2 weeks when adapted inoculum was tested.

In conclusion, nitrobenzene can be degraded under anaerobic conditions, but decomposition in general was found to be slower than under aerobic conditions.

4.5 Ultimate fate following use

During industrial processing, most nitrobenzene is retained in closed systems for use in further synthesis, predominantly of aniline, but also of substituted nitrobenzenes and anilines. Losses during production of nitrobenzene and during its subsequent use in closed systems are likely to be low; when it is used as a solvent, however, emissions are likely to be higher (e.g., US EPA, 1984) (see also sections 3.2.2 and 3.2.3).

Production losses to the environment can be estimated (see section 3.2.3), but those losses that are specifically to air are not known. Direct release of nitrobenzene to air during its manufacture can be minimized by the passage of contaminated air through activated charcoal (US EPA, 1980). Waste air at a major nitrobenzene plant in Western Europe is disposed of via a thermal treatment plant, in which nitrobenzene is converted mainly to nitrogen oxides (BUA, 1994).

Nitrobenzene waste from production and conversion and oxidation products from manufacture are decomposed thermally. At the same plant, distillation "bottoms" arising from its use as a solvent in the production of dyes are incinerated (BUA, 1994).

Because nitrobenzene is listed as a hazardous substance, disposal of waste nitrobenzene is controlled by a number of federal regulations in the USA. Land disposal restrictions apply to wastes containing nitrobenzene. These wastes may be chemically or biologically treated or incinerated by the liquid injection or fluidized bed method (HSDB, 1988; US EPA, 1988, 1989).

5. ENVIRONMENTAL LEVELS AND HUMAN EXPOSURE

5.1   Environmental levels

5.1.1   Air

Much of the information on nitrobenzene levels in air is derived from a series of reports from New Jersey, USA, in which ambient air in urban, rural and waste disposal areas was monitored extensively. In the initial study by Bozzelli et al. (1980), nitrobenzene was not detected above the level of 0.05 µg/m³ (0.01 ppb) in about 260 samples collected in 1979. In 1978, nitrobenzene levels averaged 2.0 µg/m³ (0.40 ppb) in industrial areas and 0.1 µg/m³ (0.02 ppb) and 0.46 µg/m³ (0.09 ppb) in two residential areas; in 1982, levels in residential areas were approximately 1.5 µg/m³ (0.3 ppb) or less, whereas levels in industrial areas were 46 µg/m³ (9 ppb) or more (Bozzelli & Kebbekus, 1982). Again, most of the samples were negative for nitrobenzene.

Little information is available for other areas of the USA. Pellizzari (1978) found only one positive value of 107 ng/m³ at a plant site in Louisiana. Early summarized data (US EPA, 1985) showed that less than 25% of US air samples were positive, with a median concentration of about 0.05 µg/m³ (0.01 ppb). Mean levels measured in urban areas are generally low (<1 µg/m³ [0.2 ppb]), whereas slightly higher levels (mean 2.0 µg/m³ [0.40 ppb]) have been measured in industrial areas.

Harkov et al. (1983, 1984) carried out a comparison of the concentrations of nitrobenzene at several urban sites in New Jersey, USA. In the summer, the geometric mean levels detected at three sites were 0.35, 0.35 and 0.5 µg/m³, with 80–90% of the samples being above the detection limit of 0.25 µg/m³. In contrast to this, nitrobenzene was detected in only 6–14% of the samples taken in the winter. Hunt et al. (1986), using the data collected by Harkov et al. (1984), calculated the arithmetic means for the three sites as 0.96, 1.56 and 2.1 µg/m³ in the summer and 0.050, 0.050 and 0.10 µg/m³ in the winter. In another study (Lioy et al., 1983), nitrobenzene was not detected during the winter.

Studies of air over waste disposal sites (Harkov et al., 1985) are confounded by weather and timing. Air at one landfill showed a mean nitrobenzene concentration of 6.8 µg/m³ (1.32 ppb) and another of 1.5 µg/m³ (0.3 ppb), but nitrobenzene was not detected at two other sites measured during snow and/or rain. LaRegina et al. (1986) summarized these studies by noting that the highest value for nitrobenzene was 74 µg/m³ (14.48 ppb) at a hazardous waste site, whereas nitrobenzene was often undetectable elsewhere (especially in rural areas or at sanitary landfills) or anywhere in the air during the winter.

Nitrobenzene has been detected (no level given) in forest air in Eggegebirge, Germany, and was thought to be of anthropogenic origin (Helmig et al., 1989).

Nitrobenzene was detected in 42 of 49 air samples from Japan in 1991 (detection limit 2 ng/m³). The levels measured were in the range 2.2–160 ng/m³ (Environment Agency Japan, 1992).

Some measured air levels of nitrobenzene are given in Table 8.

5.1.2 Water

5.1.2.1 Industrial and waste treatment effluents

The effluent discharge produced during nitrobenzene manufacture is the principal source of nitrobenzene release to water. Estimates of rates of losses to wastewater are discussed in section 3.2.3. The nitrobenzene in wastewater may be lost to the air, degraded by sewage organisms or, rarely, carried through to finished water.

The US Environmental Protection Agency (EPA) has surveyed nitrobenzene levels reported in effluents from 4000 publicly owned treatment works and industrial sites. The highest value in effluent was >100 mg/litre in the organic chemicals and plastics industry (Shackelford et al., 1983). Nitrobenzene was detected in 1 of 33 industrial effluents at a concentration greater than 100 µg/litre (Perry et al., 1979). Reported nitrobenzene concentrations in raw and treated industrial wastewaters from several industries range from 1.4 to

91 000 µg/litre (US EPA, 1983). Highest concentrations were associated with wastewaters from the organic chemicals and plastics industries.

Table 8. Measured levels of nitrobenzene in air

Location (samples)	Mean level (µg/m^3)[a]	Reference
Camden, USA, July–August 1981 (24-h average)	0.96 (max. 10.0)	Hunt et al., 1986
Camden, USA, January–February 1982 (24-h average)	0.050 (max. 0.75)	
Elizabeth, USA, July–August 1981 (24-h average)	1.56 (max. 24.1)	
Elizabeth, USA, January–February 1982 (24-h average)	0.050 (max. 0.35)	
Newark, USA, July–August 1981 (24-h average)	2.1 (max. 37.5)	
Newark, USA, July–August 1982 (24-h average)	0.10 (max. 1.26)	
Six sites in New Jersey, USA (sampled every 6 days for 1–2 years)	<0.050	Bozzelli & Kebbekus, 1982
Industrial site, New Jersey, USA (241 samples)	2.0	
Residential site, New Jersey, USA (49 samples)	0.10	
Residential site, New Jersey, USA (40 samples)	0.45	
Japan	0.14 (range 0.0022–0.16)	Environment Agency Japan, 1992

[a] Data are arithmetic means. Maxima or ranges are given in parentheses.

The results of two surveys of priority pollutants in publicly owned treatment works in the USA have been reported by Burns & Roe (1982). In the first survey, nitrobenzene was not detected in any influent, effluent or raw sludge samples taken from 40 publicly owned treatment works. In the second survey of 10 publicly owned treatment works, nitrobenzene was detected in 4 of 60 influent samples at 15–220 µg/litre and in 1 of 60 effluent samples at 4 µg/litre; it was not detected in any raw sludge sample.

Nitrobenzene was not detected in 238 samples of sludge taken from 204 wastewater treatment plants in Michigan, USA (Jacobs et al., 1987).

Webber & Lesage (1989) detected nitrobenzene in 27% of sludge samples taken from 15 Canadian municipal wastewater treatment plants. The median level measured was 3.5 mg/kg dry weight, and the maximum level was 9 mg/kg dry weight. Nitrobenzene is detected more frequently and at higher concentrations in effluents from industrial sources than in urban runoff. Of 1245 industrial effluents reported in the US STORET database for which analysis of nitrobenzene had been undertaken, nitrobenzene was detected in 1.8% of the samples, with the median level being <10 µg/litre (Staples et al., 1985). In the finished effluent, nitrobenzene was detected in only 3 of the 4000 publicly owned treatment works and in one oil refinery (Ellis et al., 1982). In a nationwide US project in 1982, the National Urban Runoff Program found no nitrobenzene in 86 runoff samples from 51 catchments in 19 US cities (Cole et al., 1984).

Nitrobenzene concentrations of about 20 µg/litre in the final effluent of a Los Angeles County, California, USA, municipal wastewater treatment plant in July 1978 and less than 10 µg/litre in November 1980 were reported (Young et al., 1983). Levins et al. (1979) reported only one positive sample (total sample number not stated) in Hartford, Connecticut, USA, sewage treatment plant influents, and no nitrobenzene was detected in samples taken from three other major metropolitan areas.

5.1.2.2   *Surface water*

Nitrobenzene was not detected in any surface water samples collected near 862 hazardous waste sites in the USA, according to the Contract Laboratory Program Statistical Database (CLPSD, 1988).

Nitrobenzene was not detected (detection limit 4 µg/litre) in the Potomac River, USA (Hall et al., 1987).

Detailed surveys of Japanese surface waters were undertaken in 1977 and 1991. In the 1977 survey, nitrobenzene was detected in 22 of 115 samples at a level of 0.13–3.8 µg/litre (detection limit 0.1–30 µg/litre). In the 1991 survey, nitrobenzene was detected in 1 of 153 surface water samples at a level of 0.17 µg/litre (detection limit 0.15 µg/litre). The samples were taken from both industrialized and rural areas (Kubota, 1979; Environment Agency Japan, 1992).

Staples et al. (1985) reported that of the 836 determinations of nitrobenzene in ambient surface water contained in the US STORET database, nitrobenzene was detected in 0.4% of the samples, with a median level of <10 µg/litre.

In a year-long survey of water from two reservoirs near Calgary, Canada, nitrobenzene was not detected in any of the samples taken (detection limit 0.1 µg/litre) (Hargesheimer & Lewis, 1987).

Noordsij et al. (1985) found that nitrobenzene was present in water from both the river Waal and the river Lek in the Netherlands (no levels given), but was not detected in riverbank-filtered groundwater from the same area.

In reviewing available data, generally low levels (around 0.1–1 µg/litre) of nitrobenzene have been measured in surface waters. One of the highest levels reported was 67 µg/litre, in the river Danube in Yugoslavia (Hain et al., 1990). Many of the rivers sampled for nitrobenzene are known to suffer from industrial pollution and so may not represent the general situation.

After a temporary failure in an industrial wastewater treatment plant at BASF Aktiengesellschaft in May 1984, a peak nitrobenzene concentration of 25 µg/litre was measured in the river Rhine, Germany (BUA, 1994).

Some measured levels of nitrobenzene in surface water are shown in Table 9.

5.1.2.3   *Groundwater*

Nitrobenzene was detected in groundwater at 3 of 862 hazardous waste sites in the USA at a geometric mean concentration of 1400 µg/litre, according to the Contract Laboratory Program Statistical Database (CLPSD, 1988).

Nitrobenzene was not detected (<1.13 µg/litre) in groundwater at an explosives manufacturing site in the USA. The aquifer at the site was known to be contaminated with explosives residues (Dennis et al., 1990; Wujcik et al., 1992).

Table 9. Some reported levels of nitrobenzene in surface waters

Location	Concentration (µg/litre)	Reference
River Waal, Netherlands	Spring average 5.4 Summer average 0.3 Autumn average 0.1 Winter average 1.0 Maximum level 13.8	Meijers & Van Der Leer, 1976
River Maas, Netherlands	Summer not detected (<0.1) Autumn average 0.1 Winter average 0.1 Maximum level 0.3	
Scheldt Estuary, Dutch–Belgium border	Detected in 1 sample at 0.13	Van Zoest & Van Eck, 1991
Surface water, USA	0.3–13.8	Waggot & Wheatland, 1978
River Rhine, Netherlands	Average 0.5 (1976–1978) Average 1 (1979)	Zoeteman et al., 1980
River Danube, Yugoslavia	67	Hain et al., 1990
River Rhine, Germany	Geometric mean 0.42	Kuhn & Clifford, 1986
North Saskatchewan River, Canada	Detected at 1 site at 7.25	Ongley et al., 1988
Surface water, Paris, France	0.20–0.50	Mallevialle et al., 1984
Yanghe River, China	Not detected to 0.7	Wang et al., 1999

Nitrobenzene was detected at a level of 210–250 µg/litre in groundwater from Gibbstown, USA (Rosen et al., 1992).

Bangert et al. (1988) measured nitrobenzene at a level of 4.2 mg/litre in groundwater at a coal gasification site in the USA. Nitrobenzene was not detected in groundwater at another similar site.

No nitrobenzene was detected (minimum detection limit 0.67 µg/litre) in three groundwater sources of domestic water in the Mexico City region (Downs et al., 1999).

5.1.2.4   *Drinking-water*

Nitrobenzene was detected in 1 of 14 samples of treated water in the United Kingdom. The positive sample was water derived from an upland reservoir (Fielding et al., 1981).

In a survey of 30 Canadian potable water treatment facilities, nitrobenzene was not detected in either raw or treated water (detection limit 5 µg/litre) (Otson et al., 1982).

Kopfler et al. (1977) listed nitrobenzene as one of the chemicals found in finished tap water in the USA, but did not report its concentrations or locations.

According to the BUA (1994), the nitrobenzene content in potable water following passage through the soil was 0.1 µg/litre (mean), with maximum values of 0.7 µg/litre in 50 samples taken from the river Lek at Hagestein, Netherlands, in 1986.

5.1.3   Sediments

A survey of sediment contamination in Japan was undertaken in 1991. Nitrobenzene was found in 2 of 162 sediment samples at a level of 47–70 µg/kg (detection limit 23 µg/kg) (Environment Agency Japan, 1992).

Staples et al. (1985), using the US STORET database, reported that nitrobenzene was not detected in 349 samples of sediment (detection limit 500 µg/kg dry weight).

Nitrobenzene was not detected in samples of suspended sediments or bottom sediments from the North Saskatchewan River, Canada (Ongley et al., 1988).

5.1.4   Soils

As a source of exposure of humans to nitrobenzene, soil is less important than air or groundwater. Nelson & Hites (1980) reported a nitrobenzene concentration of 8 mg/kg in the soil of a former dye manufacturing site along the bank of the industrially polluted Buffalo River in New York, USA, but failed to detect nitrobenzene in river sediments. The presence of nitrobenzene in the soils of abandoned hazardous waste sites is inferred by its presence in the atmosphere above several sites (Harkov et al., 1985; LaRegina et al., 1986). Nitrobenzene was detected in soil/sediment samples at 4 of 862 hazardous waste sites at a geometric mean concentration of 1000 µg/kg (CLPSD, 1988).

Nitrobenzene was detected at a level of 0.79 mg/kg in 1 of 10 soil samples taken from a site in Seattle, USA. The site had formerly been used for coal and oil gasification, and it was thought that many wastes were deposited on the site (Turney & Goerlitz, 1990).

5.1.5 Food

Data on nitrobenzene occurrence in foods were not found in the literature. Nitrobenzene has been detected as a bioaccumulated material in fish samples in Japan (see section 5.1.6). No monitoring of plant tissues has been reported, even though uptake of nitrobenzene by plants has been observed (see section 4.3.2).

5.1.6 Occurrence in biota

Surveys of nitrobenzene in fish were carried out in Japan in 1991. Nitrobenzene was detected in 4 of 147 fish samples at a level of 11–26 µg/kg (detection limit 8.7 µg/kg) (Environment Agency Japan, 1992).

Staples et al. (1985), using the US STORET database, reported that nitrobenzene was not detected in 122 biota samples (detection limit 2 mg/kg wet weight).

5.2 General population exposure

General exposure of the population to nitrobenzene is limited to variable concentrations in air and possibly drinking-water (see section 5.1). There is likely to be very limited exposure via the use of nitrobenzene in some consumer products, although the potential for exposure from these uses has not been quantified (see sections 3.3 and 5.1). Urban areas are likely to have higher levels of nitrobenzene in air in the summer than in the winter due to both the formation of nitrobenzene by nitration of benzene (from motor vehicle fuels) and the higher volatility of nitrobenzene during the warmer months (Dorigan

& Hushon, 1976; see also sections 3.2.4 and 5.1). Ambient exposure in the winter may be negligible.

Based on air studies and on estimates of releases during manufacture, only populations in the vicinity of manufacturing activities (i.e., producers and industrial consumers of nitrobenzene for subsequent synthesis) and petroleum refining plants are likely to have any significant exposure to nitrobenzene (see section 5.1). However, consideration of possible groundwater (see section 5.1.2) and soil contamination (see section 5.1.4) and uptake of nitrobenzene by plants (see section 4.3.2) expands the population with the potential for higher exposures to those people living in and around abandoned hazardous waste sites.

5.3 Occupational exposure during manufacture, formulation or use

Occupational exposure is likely to be significantly higher than the exposure of the general population. In the USA, a NIOSH (1990) survey estimated that about 10 600 workers (mainly chemists, equipment servicers and janitorial staff) may be potentially exposed in facilities in which nitrobenzene is used; this survey was non-quantitative with respect to the amount of exposure. Because nitrobenzene is readily absorbed through the skin, skin contact may be a major, if not the main, route of exposure under occupational circumstances. It was experimentally found (see section 6.1.3) that uptake of nitrobenzene vapour at 5 mg/m^3 through the skin or the whole body area amounted to 7 mg during 6 h, whereas the corresponding uptake via inhalation can be estimated as about 30 mg.

In the 1950s, air measurements of nitrobenzene in a Hungarian production facility gave 29 mg/m^3 (confidence interval 20–38 mg/m^3), and the workers showed increased methaemoglobin levels (mean 0.61 g/100 ml, i.e., about 4–5%), some workers even indicating Heinz body formation at about 1% (Pacséri et al., 1958). In a more recent survey in an anthraquinone plant (Harmer et al., 1989), atmospheric nitrobenzene levels over an 8-h period, measured by GC, ranged from 0.7 to 2.2 mg/m^3. Small amounts of unchanged nitrobenzene were detected in blood preshift on day 1 (ranging from non-detectable to 52 µg/litre) and postshift on day 3 (ranging from 20 to 110 µg/litre in the same seven workers), indicating some accumulation of nitrobenzene in the body. Methaemoglobin levels were all below 2%, with no clear correlation with blood nitrobenzene levels. Urinary *p*-nitrophenol tended to increase over the 3-day shift period (ranging between about 0.2 and 5.4 mg/litre), although there did not appear to be much correlation with atmospheric concentrations of nitrobenzene. Among many possible factors, the lack of correlation could suggest skin absorption. Urinary *m*- and *o*- nitrophenols were "detected."

6. KINETICS AND METABOLISM IN LABORATORY ANIMALS AND HUMANS

6.1 Absorption

Nitrobenzene is readily taken up via ingestion, dermal absorption and inhalation.

6.1.1 Oral exposure

After oral administration of 250 mg [^{14}C]nitrobenzene/kg of body weight by stomach tube to a rabbit, Parke (1956) recovered 78% of administered radioactivity within 8 days; 1.2% was in exhaled air, 57.6% in urine, 12.2% in faeces and gastrointestinal tract contents, and 7.5% in tissues, of which fat deposits had the highest levels. This would indicate that the extent of gastrointestinal absorption was at least 66% of the oral dose, assuming no enterohepatic cycling.

In rats (Fischer-344 and Sprague-Dawley strains) given single oral doses of 22.5 or 225 mg [^{14}C]nitrobenzene/kg of body weight, approximately 72–88% of the administered dose was recovered in 72 h, of which about 80% was in urine, indicating that there was significant absorption of nitrobenzene from the gastrointestinal tract (Rickert et al., 1983). Oral doses appeared to be less well absorbed in B6C3F$_1$ mice than in the two rat strains; at 72 h after an oral dose of 225 mg/kg of body weight, about 35% was found in urine (compared with 57–63% in rats at the same dose) and about 19% in the faeces (compared with 14% in rats).

No studies were located regarding the extent of uptake of nitrobenzene by humans after oral exposure; however, from reports of human poisonings (see section 8.1.1), oral absorption would appear to be rapid and extensive.

6.1.2 Dermal exposure

The toxicokinetics of dermal exposure have not been well studied in either experimental animals or humans. In animal studies, nitrobenzene appears to be well absorbed after dermal application, based on observations of toxic responses and pathological findings in treated animals (Shimkin, 1939; Matsumaru & Yoshida, 1959). Similarly, numerous poisonings, with some fatalities, have occurred in humans dermally exposed to nitrobenzene (see section 8.1.2).

An apparatus was developed to measure skin absorption from nitrobenzene vapour in the air, without inhalation of nitrobenzene (Piotrowski, 1967). The author calculated that approximately half as much nitrobenzene vapour was absorbed through the skin as through the lungs when volunteers were exposed to 5–30 mg/m^3. Vapour absorption through the skin was proportional to the concentration of nitrobenzene in the air, and normal working clothes reduced this absorption by only 20–30%. In high humidity, skin absorption of vapour was significantly increased. It was estimated that at an air level of 5 mg/m^3, intake from dermal exposure to vapour would amount to about 7 mg over a 6-h working day. Assuming lung ventilation of 4–5 m^3 over a 6-h period and 80% retention, lung uptake would be 18 mg, for a total intake of 25 mg/day. It was noted that whereas skin absorption from vapour exposure was substantial (absorption rate per unit vapour concentration between 0.23 and 0.3 mg/h per mg/m^3 over the concentration range 5–30 mg/m^3), it was much less than the absorption of liquid nitrobenzene through the skin, which can reach about 2 mg/cm^2 per hour (Salmowa & Piotrowski, 1960).

The Task Group noted that a more realistic lung ventilation of 7.5 m^3 over 6 h would lead to an estimated lung uptake of 30 mg (assuming 80% retention), in which case vapour absorption through the entire skin would be only about one-fifth to one-quarter of that through the lungs.

Feldmann & Maibach (1970) applied 4 $\mu g/cm^2$ of a number of ^{14}C-labelled organic compounds, including nitrobenzene, to a 13-cm^2 circular area of the forearm of one human volunteer for 24 h (sites unprotected, but not washed for at least 24 h). Radioactivity in urine was then measured over a 5-day collection period (collection of carbon dioxide after wet ashing); compounds were dissolved in acetone for application. Total absorption of nitrobenzene was estimated to be only 1.53% of the total dose under the above conditions. In view of the fact that only a small amount of nitrobenzene was applied (52 μg) in acetone solvent to an unprotected site, the finding of only limited absorption must be interpreted with caution.

6.1.3   Inhalation exposure

Five adult Wistar rats were exposed to nitrobenzene vapour at 130 mg/m³ (25 ppm) for 8 h, and the urine was collected for three 24-h periods. No quantitative estimation of absorption was made, but both *p*-nitrophenol and *p*-aminophenol were detected in the urine at the 24- and 48-h collection periods. In 24-h urine, levels of the former were about 1.7 μmol/24 h per rat, and levels of the latter were about 1.1 μmol/24 h per rat; both declined to less than one-half these amounts at the 48-h collection (Ikeda & Kita, 1964).

In humans, nitrobenzene is well absorbed through the lungs (WHO, 1986). During a 6-h exposure of volunteers to nitrobenzene, Salmowa et al. (1963) found absorption to average 80% (73–87%) in seven men breathing 5–30 mg nitrobenzene/m³. Uptake was dose dependent but showed considerable interindividual variation.

6.2   Distribution

6.2.1   Oral exposure

Male Wistar rats were orally dosed (stomach tube) with ¹⁴C-labelled nitrobenzene at 25 μg/rat per day for 3 days. Faeces and urine were collected every day until necropsy on day 8, with determination of radioactivity in selected tissues (Freitag et al., 1982). Seven-day excretion accounted for 59.3% of the dose in the urine and 15.4% in the faeces. At necropsy, no detectable radioactivity was found in abdominal adipose tissue or in the lungs, whereas 0.43% of the dose was in the liver and 2.3% was retained in the remaining carcass. Not all sources of elimination of label were measured (e.g., carbon dioxide in expired air), and 22.6% of the administered dose was not accounted for.

In a study in female Wistar rats that received [U-¹⁴C]nitrobenzene by gavage at a dose of 0.2 mmol/kg of body weight, tissue concentrations of total radioactivity were determined after 1 and 7 days. Within 24 h, 50 ± 10% of the radioactive dose appeared in the urine and about 4% in the faeces. After 1 week, 65% of the dose had appeared in the urine and 15.5% in the faeces. Binding to tissues after the first day, in (pmol/mg)/(mmol/kg of body weight), was as follows: blood, 229; liver, 129; kidney, 204; and lung, 62. At day 7, these same tissue levels were 134, 26.5, 48 and 29 (pmol/mg)/(mmol/kg of body weight). Binding to haemoglobin (Hb) was also studied (Albrecht & Neumann, 1985) (see section 6.6).

Analysis of radioactivity in body tissues of one rabbit indicated that 1.5 days after dosing with 250 mg [^{14}C]nitrobenzene/kg of body weight by stomach tube, 44.5% of the administered dose was accounted for in tissues (excluding stomach and intestinal contents), particularly in kidney fat (15.4%), skeletal muscle (12%) and intestinal fat (11.6%); unchanged nitrobenzene was present in the tissues. Radioactivity in the same tissues was down to <7.5% of the applied dose after 8 days, with no unchanged nitrobenzene detected (Parke, 1956).

In autopsies of five cases of nitrobenzene poisoning in humans, the chemical was found in stomach, liver, brain and blood. The highest concentration found in liver was 124 mg/kg tissue, and in brain, 164 mg/kg tissue (Wirtschafter & Wolpaw, 1944).

6.2.2 Dermal exposure

No studies were located that investigated the distribution of nitrobenzene or its metabolites after dermal exposure.

6.2.3 Inhalation exposure

No studies on the distribution of nitrobenzene or its metabolites after inhalation exposure of animals or humans were found in the literature.

6.3 Metabolic transformation

6.3.1 In vivo *metabolic studies*

Following on from a review of nitrobenzene metabolism by Beauchamp et al. (1982), which covered articles published prior to January 1981, Rickert (1987) reviewed subsequent papers, relying heavily on the work of Levin & Dent (1982). A metabolic diagram arising from that review is presented in Figure 3. A more recent review by Holder (1999a) covers the persistence of reactive intermediates of nitrobenzene metabolism in tissues.

Most metabolic studies have utilized oral dosing; thus, it is not possible to determine whether there are quantitative or qualitative differences in nitrobenzene metabolism following oral, dermal and inhalational exposure.

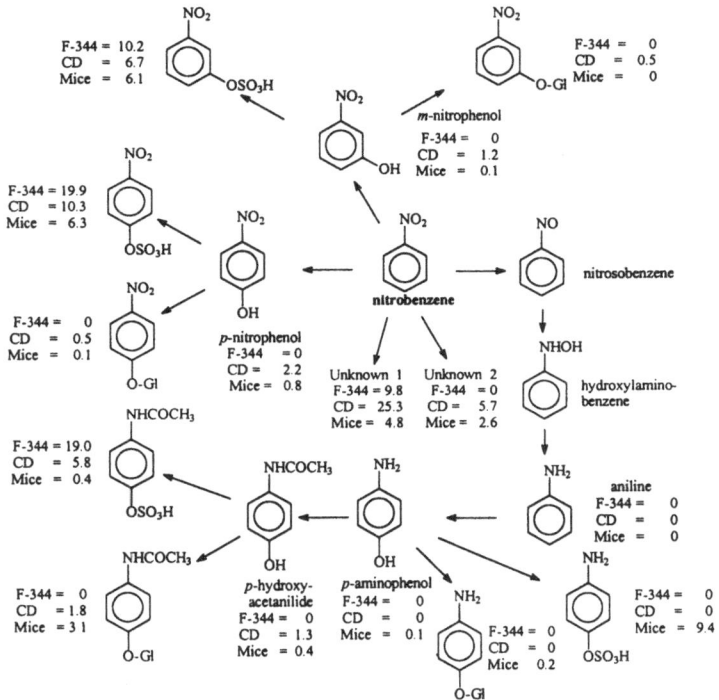

Fig. 3. Urinary metabolites of nitrobenzene in Fischer-344 and CD rats and B6C3F₁ mice. Both reduction and oxidation pathways for nitrobenzene are shown. Numbers beside the structures are the percentages of a 225 mg/kg of body weight oral dose found in the urine in 72 h (Rickert, 1987). Note that hydroxylaminobenzene = N-phenylhydroxylamine. Abbreviations: F-344 — Fischer-344 rats; CD — CD rats; mice — B6C3F₁ mice; Gl — glucuronic acid.

Robinson et al. (1951) found that in giant chinchilla rabbits, nitrobenzene was both oxidized and reduced. In semiquantitative studies (using selective extractions, chromatography and colorimetric assays), urine of female giant chinchilla rabbits was collected for 48 h after gavage dosing (via stomach tube) with about 200 mg nitrobenzene/kg of body weight. The major metabolite was *p*-aminophenol

(35% of the dose), followed by *p*- and *m*-nitrophenol (5% and 4%, respectively) and traces of aniline (0.5%), *o*- and *m*-aminophenol (0.54% and 0.58%, respectively), 4-nitrocatechol (0.5%) and *o*-nitrophenol (0.05%); unchanged nitrobenzene constituted only 0.03% of the dose recovered after 2 days. Over 48 h, just over 50% of the dose was accounted for; metabolites could still be detected in urine 7 days after dosing. The *p*-aminophenol was excreted as glucuronic and sulfuric acid conjugates. Since the amount of nitrobenzene excreted unchanged was negligible, virtually the whole of the dose was either oxidized (nitrophenols) or reduced (aniline and aminophenols); roughly two-thirds was excreted in the reduced form as amino compounds, and one-third as nitro compounds.

The work cited in the previous paragraph was followed up using radioactive nitrobenzene, leading to the identification of a much greater proportion of the orally administered dose (Parke, 1956). In rabbits given a single dose of randomly labelled [^{14}C]nitrobenzene (200–400 mg/kg of body weight) by stomach tube, 70% of the radioactivity was eliminated from the animals in expired air, urine and faeces within 4–5 days after dosing. The remainder of the radioactivity was found to be slowly excreted in urine and possibly in expired air (as carbon dioxide). Approximately 1% of the dose was found as carbon dioxide and 0.6% as nitrobenzene in expired air within the first 30 h of exposure; the elimination of carbon dioxide was not complete within this time. After 4–5 days, approximately 58% of the dose was eliminated in urine as *p*-aminophenol (31%), *m*-nitrophenol (9%), *p*-nitrophenol (9%), *m*-aminophenol (4%), *o*-aminophenol (3%), 4-nitrocatechol (0.7%), aniline (0.3%) and *o*-nitrophenol (0.1%); two new metabolites, not identified in the studies of Robinson et al. (1951), were nitroquinol (0.1%) and *p*-nitrophenolmercapturic acid (0.3%). Over 4–5 days after dosing, faeces were found to contain around 9% of the administered radioactivity, of which 6% was as *p*-aminophenol. Over this period, including the radioactivity remaining in the tissues and that excreted in urine, faeces and expired air, 85–90% of the dose was accounted for. Traces of radioactivity were still detectable in urine 10 days after dosing (Parke, 1956).

In a study in which a single guinea-pig was given [^{14}C]nitrobenzene by intraperitoneal injection at 500 mg/kg of body weight (Parke, 1956), the only marked differences in the findings from the results in rabbits (gavage dosing) were the complete absence of *o*-

nitrophenol and a decrease in *m*-nitrophenol; it was considered by the author that the *o*- and *m*-nitrophenols may be more readily reduced to the corresponding aminophenols in guinea-pigs than in rabbits.

Rickert et al. (1983) studied the metabolism and excretion of orally administered nitrobenzene in B6C3F$_1$ mice, Fischer-344 rats and Sprague-Dawley (CD) rats. Findings are summarized in Figure 3 (taken from Rickert, 1987). In mice and both rat strains, urinary metabolites were free and conjugated forms of *p*-hydroxyacetanilide, *p*- and *m*-nitrophenol and, in mice only, *p*-aminophenol; Fischer-344 rats excreted these exclusively as sulfate esters, and mice and CD rats, as free compounds, sulfate esters and glucuronides.

When Wistar rats were given 25 mg radiolabelled nitrobenzene/kg of body weight by gavage, biotransformation was first seen in the intestine, where nitrobenzene was sequentially reduced to nitroso-benzene, hydroxylaminobenzene and aniline (Albrecht & Neumann, 1985). These findings were also reported in Fischer-344 rats (Levin & Dent, 1982).

The action of bacteria normally present in the small intestine of the rat is an important element in the formation of methaemoglobin resulting from nitrobenzene exposure. Germ-free rats do not develop methaemoglobinaemia when intraperitoneally dosed with nitrobenzene (Reddy et al., 1976). When nitrobenzene (200 mg/kg of body weight in sesame oil) was intraperitoneally administered to normal Sprague-Dawley rats, 30–40% of the haemoglobin in the blood was converted to methaemoglobin within 1–2 h. When the same dose was adminis-tered to germ-free or antibiotic-pretreated rats, there was no measur-able methaemoglobin formation, even when measured up to 7 h after treatment. The nitroreductase activities of various tissues (liver, kid-ney, gut wall) were not significantly different in germ-free and control rats, but the activity was negligible in gut contents from germ-free rats and high in control rats. This led the authors to suggest that a nitro-benzene metabolite such as aniline (which is formed by the bacterial reduction of nitrobenzene in the intestines of rats) is involved in met-haemoglobin formation. In addition, diet has been shown to play a role in the production of methaemoglobin by influencing the intestinal microflora; the presence of pectin in the diets of rats was shown to increase the ability of orally administered nitrobenzene to induce met-haemoglobinaemia. This was correlated with the increased *in vitro* reductive metabolism of [^{14}C]nitrobenzene by the caecal contents of rats fed purified diets containing increasing amounts of pectin (Goldstein et al., 1984b).

Confirming and extending the results of Reddy et al. (1976), Levin & Dent (1982) studied the metabolism of [U-^{14}C]nitrobenzene in Fischer-344 rats, both *in vivo* and by hepatic microsomal fractions and caecal contents. Isolated caecal contents sequentially reduced nitrobenzene to nitrosobenzene, hydroxylaminobenzene and, ultimately, aniline; this anaerobic metabolism occurred 150 times faster than reduction by microsomes, even when the microsomes were incubated under optimal anaerobic conditions (which are unlikely to occur *in vivo*). After oral dosing of rats, final reduction products in the urine (of which the major one was *p*-hydroxyacetanilide) arise from these absorbed gut flora metabolites, which undergo subsequent hydroxylations and acetylations, presumably by the mammalian mixed-function oxidases and acetyltransferases. *In vivo* antibiotic treatment reduced urinary excretion of *p*-hydroxyacetanilide and another unidentified reduced metabolite by 94% and 86%, respectively, indicating that whereas gut microflora was the major cause of reductive metabolism, some limited reductive metabolism may occur in liver, or that the antibiotic treatment did not eliminate all of the reductive bacteria from the gut.

Further proof of enteric nitroreduction in rats is the demonstration that a model compound such as *p*-nitrobenzoic acid is reduced to *p*-aminobenzoic acid in normal rats (25% occurring in urine), but not significantly so in germ-free animals (1% in urine) (Wheeler et al., 1975). Selective introduction of various gut bacteria such as *Clostridium* and *Streptococcus faecalis* into germ-free rats increased the reduction of *p*-nitrobenzoic acid to *p*-aminobenzoic acid from negligible levels to around 12%. Removal of a substantial portion of the caecum also decreased the capacity of normal rats to reduce *p*-nitrobenzoic acid.

Oxidative metabolism proceeded independently of reductive metabolism in this study; however, the rate of oxidative metabolism of nitrobenzene by hepatic microsomes was extremely slow, being about one-tenth that of aniline. The limited role of the mixed-function oxidase activity for nitrobenzene may be a consequence of the uncoupling of cytochrome P-450, resulting in futile redox cycling of the nitro group, with the generation of superoxide anions. Major oxidative metabolites, *m*- and *p*-nitrophenol, represented >30% of the dose. Since rates of hepatic and extrahepatic (renal) microsomal metabolism or isolated hepatocyte metabolism did not appear to account for this *in vivo* level of nitrophenol formation, it was not possible to identify the subcellular site(s) of oxidative metabolism. The fact that oxidative metabolism was unaffected by antibiotic treatment indicates that the *m*- and *p*-nitrophenol did not arise from ring hydroxylation of reduced metabolites followed by reoxidation of the nitrogen. Reductive metabolism appears to precede oxidative metabolism, for two reasons. First, since at least 30% of the nitrophenol was excreted in the *meta* form, if reduction were secondary to oxidation, it would be expected that some of the reduction products would be in the *meta* form, and this was not the case. Second, the sulfated and acetylated nature of the final reduction product implies liver-catalysed final steps following reduction in the gut (Levin & Dent, 1982). An outline of the proposed metabolism is given in Figure 3 above.

6.3.2 In vitro *and* ex vivo *metabolic studies*

Sheep ruminal content was able to reduce nitro groups of nitro-benzene (Acosta de Pérez et al., 1992).

In rat liver microsomes, *p*-nitrophenol, an intermediate of nitro-benzene, can be enzymatically converted to 4-nitrocatechol (Chrastil & Wilson, 1975; Billings, 1985). The presence of a highly active *p*-nitrophenol hydroxylase catalysing the formation of 4-nitrocatechol from *p*-nitrophenol (by placing a hydroxy group at the *ortho* position) was detected in sheep lung microsomes (Arinç & Aydoğmus, 1990). The formation of catechols from benzene and nitrobenzene has been implicated in the possible carcinogenic activity of these compounds (Billings, 1985; Kalf et al., 1987); catechols are chemically reactive and may covalently bind to cellular macromolecules.

Daly et al. (1968) studied the metabolism of nitrobenzene in female guinea-pigs *in vivo* and in rat liver microsomes *in vitro*. Animals were given 400 mg nitrobenzene/kg of body weight intra-peritoneally, and the urine was collected for 48 h. The main metabolite found was 4-hydroxynitrobenzene (*p*-nitrophenol). Traces of this metabolite were found when nitrobenzene was incubated at 37 °C with rat liver microsomes.

Male Wistar rats were given subcutaneous doses of 150 mg nitro-benzene/kg of body weight per day for 3 days or 5 or 50 mg/kg of body weight per day for 30 days. There was 34–60% stimulation of nitrobenzene reductase activity and 17–47% stimulation of aniline *p*-hydroxylase activity in 9000 × *g* supernatants of livers from these animals (Wišniewska-Knypl et al., 1975). This suggests that repeated exposure to nitrobenzene may have some effects on the rate or route of its own metabolism.

In an English abstract of a Russian paper, it was reported that electron spin resonance (ESR) studies revealed that the reduction of nitrobenzene in liver homogenates is accompanied by the appearance of nitroanion radicals. The ESR spectrum of nitroxyl radicals of nitrobenzene was recorded. The formation of haemoprotein nitrosyl complexes was absent for nitrobenzene and nitrosobenzene (Baider & Isichenko, 1990). This result would tend to confirm the results of Levin & Dent (1982) (see section 7.9.1).

The metabolism of the aromatic compounds aniline, acetanilide, *N*-hydroxyacetanilide, nitrosobenzene and nitrobenzene was investi-gated in boar spermatozoa fortified with glucose. No acetylation, deacetylation or mono-oxygenation of these compounds was found. Nitrobenzene was reduced slowly, with the formation of trace amounts of hydroxylaminobenzene. Nitrosobenzene was a good substrate for this reductive reaction; the products were *N*-hydroxyacetanilide, azoxybenzene and an organic-phase non-extractable metabolite (Yoshioka et al., 1989).

6.3.3 *Biochemical and mechanistic considerations*

Although nitrobenzene undergoes more C-hydroxylation (ring oxidation to produce phenols) than *N*-hydroxylation (nitroxides), the latter path is important because toxic and potentially carcinogenic nitroxide intermediates can be formed (Kiese, 1966; Miller, 1970; Weisberger & Weisberger, 1973; Mason, 1982; Blaauboer & Van Holsteijn, 1983; Verna et al., 1996). Nitroarenes such as nitrobenzene can be metabolized at the nitro group to the same nitroxide inter-mediates as their analogue aromatic amines (Blaauboer & Van Holsteijn, 1983). For example, nitrobenzene can be *N*-reduced to form nitrosobenzene and phenylhydroxylamine (Figure 4A). Conversely, aniline (either administered as such or formed from nitrobenzene) can

A. Nitrobenzene bacterial reduction mechanism (caecum)

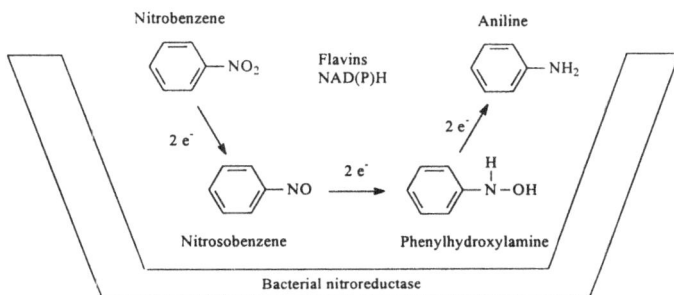

Fig. 4. Oxidation and reduction mechanisms for nitrobenzene.
A. *Microbial nitroreductase reaction in the intestines of rats* — This
appears to be the most important nitroreduction site in the body following
oral exposure to nitrobenzene. Note that reactive intermediates
nitrosobenzene, phenylhydroxylamine and aniline are released and then
absorbed and distributed in the body. Free radicals are not released
locally from the nitroreductase catalytic centre (Holder, 1999a).

be *N*-oxidized to form both of these intermediates. Acetylation of
phenylhydroxylamine at the hydroxylamine group is known to occur
by polymorphic acetyltransferases in rodents and humans (and the
chemically reactive product, *N*-hydroxyacetanilide, may possess
carcinogenic potential) (Hein, 1988; Hein et al., 1997).

Intestinal nitroreduction of nitroarenes is an important nitroaromatic metabolic pathway. Cellular alterations are not likely to be
limited to DNA. Reducing metabolism can occur by caecal bacteria
that contain different nitroreductases, some oxygen sensitive, some not
(Wheeler et al., 1975; Peterson et al., 1979). For example, Figure 4A
outlines the metabolic action of an enteric microbial nitroreductase, an
NAD(P)H-dependent flavoprotein monomer (27 kilodaltons) insensitive to oxygen (Bryant & DeLuca, 1991). The active centre catalyses
the reaction $R-NO_2$ to $R-NH_2$ by a three-step, two-electron per step
mechanism. The enteric metabolic site was identified by observations
that either antibiotic-pretreated or germ-free rats have much lower
amounts of both enteric bacteria and caecal nitroreduction activity.
Conversely, caecal nitroreduction increases in germ-free rats that are
inoculated intestinally with normal rat caecal lavage (Wheeler et al.,
1975; Levin & Dent, 1982; see section 6.3.1). Although reduction by
gut bacteria is the primary reductive mechanism for orally

administered nitrobenzene, small amounts of inhaled nitrobenzene may be swallowed and also undergo reduction in the intestines.

B. Nitrobenzene microsomal reduction mechanism

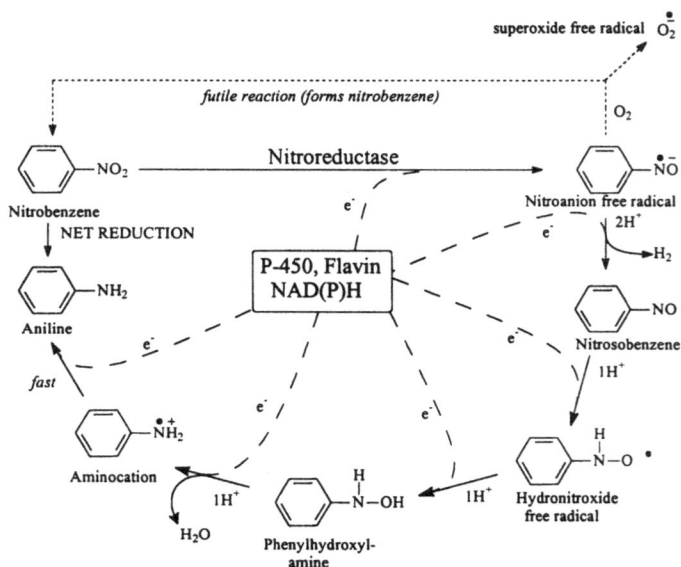

Fig. 4 (Contd). Oxidation and reduction mechanisms for nitrobenzene. **B.** *Microsomal one-electron per step mechanism for nitrobenzene reduction* — Note the cellular production of reactive free radicals: superoxide, nitroanion, hydronitroxide and the aminocation. At fast rates of nitrobenzene catalysis, these free radicals may exceed the capacity of the local tissue to quench these reactive species by spin traps. Although all these reactions apparently are driven by microsomal P-450s, NAD(P)H pools and perhaps mitochondrial flavins, these redox mechanisms are not well understood. The aminocation free radical, although diagrammed here, has not been isolated and remains a theoretical chemical supposition (Holder, 1999a).

Nitrosobenzene and phenylhydroxylamine are released in the gut after nitroreduction of nitrobenzene (Figure 4A), followed by systemic absorption of these metabolites plus parent compound. These compounds then undergo further systemic metabolism, e.g., hydroxylation and acetylation, presumably by mixed-function oxidases and acetyltransferases (Levin & Dent, 1982), as well as further reduction by tissue microsomal enzymes; the cellular microsomal one electron per step reduction sequence acts on nitrobenzene systemically absorbed by all routes of exposure (Figure 4B).

As reported above (section 6.3.1), the anaerobic kinetics of rat caecal nitroreductases were measured to be 150 times faster than the hepatic microsomal nitroreductase kinetics (Reddy et al., 1976; Levin & Dent, 1982). This would suggest that a significant portion of nitrobenzene reduction takes place primarily in the gut. However, further consideration suggests that there are many more tissue nitroreductase sites than in the gut and that nitrobenzene actually dwells in tissues for longer. Thus, notwithstanding the observation that tissue microsomal nitroreductase rates are considerably slower than the gut bacterial rates, body tissues and organs are likely to reduce a measurable portion of a nitrobenzene dose over time, although the amount relative to that occurring in the gut is unknown. It is likely that the initial systemic exposure to nitrobenzene reduction products would arise from gut reduction mechanisms, whereas later-stage exposures to reduced metabolites could arise from tissue microsomal metabolism, especially in the liver. The implications of this for nitrobenzene's toxicokinetics and toxicodynamics are not clear at this time (for further discussion, see Rickert et al., 1983; Rickert, 1984, 1987).

Microsomal oxidative metabolism of nitrobenzene, which forms various phenols, proceeds slowly compared with the *N*-reduction of nitrobenzene (see section 6.3.1). That is, not only are the bacterial gut nitroreductases faster than systemic tissue microsomal nitroreductases (by about 150-fold on a per gram tissue basis), microsomal nitroreductases are faster than microsomal oxidation processes for nitrobenzene (by about 15-fold on a per milligram microsomal protein basis). Slow ring oxidation may contribute to slow overall nitrobenzene metabolism (Levin & Dent, 1982).

Figure 4B illustrates the systemic microsomal reductive pathway for nitrobenzene, with the formation of a nitroanion free radical (Fouts & Brodie, 1957). NADPH mediates the abstraction of one electron from nitrobenzene, and the nitroanion free radical forms (Mason & Holtzman, 1975a). The nitroanion free radical generated from nitrobenzene has a sufficiently long half-life of 1–10 s that it can be detected by emission of a unique ESR spectrum. The more extended the ring aromaticity in nitroaromatic compounds, the longer the free radical half-life that is expected. This is based on electronic delocalization, which leads to radical longevity. Nitroanion free radicals have sufficient longevity to travel and react; the longer the residence time, the more chances they have to react with cellular macromolecules. The presence of all four free radicals shown in Figure 4B can be detected non-destructively by their specific ESR signal patterns (Mason & Holtzman, 1975a; Maples et al., 1990).

Exposure to a number of nitroaromatic compounds, including nitrobenzene, increases tissue oxygen uptake (Sealy et al., 1978). As illustrated in Figure 4B, nitrobenzene is reduced by a microsomal reductase to form the nitroanion free radical and, instead of being further reduced, can react with tissue oxygen to reform nitrobenzene and create the superoxide free radical, $O_2^{\cdot-}$ (Bus & Gibson, 1982; Bus, 1983). A number of chemicals undergo oxygen-related cycling involving reduced intermediates, with reformation of the parent chemical (e.g., Klaassen, 2001). By spin-trapping with phenyl-*tert*-butyl nitrone, Sealy et al. (1978) showed that the superoxide radical was produced. Mason & Holtzman (1975b) observed that the catabolic enzymes superoxide dismutase and catalase can lower nitroaromatic-induced oxygen consumption. Superoxide dismutase mediates a dismutation reaction of the superoxide free radicals: $2O_2^{\cdot-} \rightarrow O_2 + 2H_2O_2^{\cdot}$ (Heukelekian & Rand, 1955; Mason & Holtzman, 1975b; Flohé et al., 1985). Catalase catalyses the rapid decomposition of hydrogen peroxide: $2H_2O_2^{\cdot} \rightarrow O_2 + 2H_2O$. The superoxide free radical that is generated after nitroaromatic exposure dismutates to form oxygen and hydrogen peroxide, which, in turn, is dissociated by catalase to form oxygen and water. Notably, both reactions replace some of the nitroaromatic-induced oxygen consumption. If significant nitrobenzene exposure takes place, nitroanion free radicals continue to react with tissue oxygen to form accumulating $O_2^{\cdot-}$ in what is known as a futile loop reaction, so designated because nitrobenzene is continually reproduced and hence futilely metabolized (Mason & Holtzman, 1975b; Sealy et al., 1978). Generation of reactive radicals or oxygen species (i.e., nitroanion free radicals, superoxide anion free radical, hydrogen peroxide) can disturb the redox balance of cells (Trush & Kensler, 1991). The disturbance in the balance of oxygen-reactive species and their expeditious elimination can lead to oxidative stress (Gutteridge, 1995). In particular, the superoxide free radical is known to be quite reactive and toxic (Kensler et al., 1989; Keher, 1993; Dreher & Junod, 1996).

The intermediates shown in Figure 4B are chemically reactive and undergo a number of side-reactions that are not shown, but which could nonetheless be important in nitrobenzene disposition and active metabolite formation. Not only can the nitroanion free radical react with oxygen (above), but it also can self-react in a disproportionation reaction, thereby reproducing nitrobenzene as a product plus nitroso-benzene (Mason & Holtzman, 1975b; Sealy et al., 1978; Levin et al., 1982). Again, with 1 mol of nitrobenzene being reformed for every 2 mol nitroanion free radical reacting, there is a sparing action to the completion of nitrobenzene metabolism, just as is the case for the futile loop reaction. The intermediates nitrosobenzene and phenyl-hydroxylamine can undergo a condensation reaction to form azoxy-benzene: $PhNO + PhNHOH \rightarrow PhN=N^+(O^-)Ph + H_2O$ (where Ph is the phenyl moiety) (Pizzolatti & Yunes, 1990). The latter reaction is supported by acid or base catalysis in water. It seems likely that there may be a background of non-enzymatically converted PhNO to $PhN=N^+(O^-)Ph$ (Corbett & Chipko, 1977).

Because of the ubiquity of the redox conditions capable of pro-ducing nitrosobenzene and phenylhydroxylamine in the gut and in various organs, a number of tissues can be damaged — and, because of free radical chain reactions, not necessarily where the free radicals are originally produced (Holder, 1999a). A nitroxide intermediate radical can pass off the unpaired electron to an acceptor molecule, which can act as a carrier. This new radical can, if stable enough, pass the electron to yet another acceptor, and the process propagates in space and time. A number of factors in each tissue can affect the number of free radical producers, quenching agents and spin traps that can act as stabilizers, inhibitors and carriers for free radical transport (Stier et al., 1980; Keher, 1993; Gutteridge, 1995; Netke et al., 1997). These factors, in turn, may influence the toxicity profile of nitro-benzene in different organs and tissues.

Perfused liver studies show that added aniline, nitrosobenzene or phenylhydroxylamine produce only modest ESR signals. Due to ample liver reductive capability, the nitroxide intermediates are readily reduced to aniline, leaving only small steady-state quantities of nitro-sobenzene and phenylhydroxylamine and associated free radicals in the liver that produce the specific signal (Kadlubar & Ziegler, 1974; Eyer et al., 1980). There is suggestive evidence of the uncoupling of liver P-450 complex if nitrobenzene exposure is sufficiently high (Levin et al., 1982).

The apparent limited role of mixed-function oxidase metabolic activities for nitrobenzene over time may be a consequence of the uncoupling of the cytochrome P-450 complex, which can lead to "futile redox cycling" of the nitro group, with the continual generation of superoxide anions (Bus & Gibson, 1982; Levin et al., 1982; Bus, 1983; Rickert, 1987; see also section 7.9.5). Because neither extrahepatic (renal) microsomal rates nor isolated hepatocyte microsomal metabolism appeared to be able to account for the *in vivo* level of nitrophenol formation, it was not possible from the Levin & Dent (1982) study to identify all the subcellular sites of oxidative metabolism; the mitochondria might be the site of the greater amount of nitrophenol formation, but this is unknown.

6.3.4   *Metabolism in erythrocytes*

Aniline formed by the reduction of nitrobenzene was originally considered to be responsible for methaemoglobin formation following nitrobenzene exposure. Others examined the role of phenylhydroxylamine in causing Fe^{2+}-haemoglobin to be oxidized to Fe^{3+}-haemoglobin in red blood cells (Smith et al., 1967). Later, it was suggested that oxidative damage to red cells may arise from hydrogen peroxide formed as a result of "auto-oxidation" of quinone intermediates, such as *p*-aminophenol (Kiese, 1966). This consideration was based on the fact that *p*-aminophenol itself, given *in vivo*, produces methaemoglobin. In addition, superoxide free radicals are generated in a futile reaction cycle during the metabolism of nitrobenzene or other nitroaromatic chemicals (e.g., Klaassen, 2001); the parent nitro or nitroxide compound that enters red blood cells is reformed in a futile redox cycle (Levin & Dent, 1982; see also section 6.3.3). In relation to the damage that superoxide anions may cause red blood cells, it is known that superoxide dismutase and methaemoglobin reductase are essential enzymes in the oxidant protection of erythrocytes (Luke & Betton, 1987).

Although the redox chemistry (discussed above) is applicable throughout the body, special attention has been given to red blood cells because of the specific redox chemistry taking place there. In the nitrobenzene reduction sequence, nitrosobenzene is the first stable chemical formed, which, in turn, is reduced to phenylhydroxylamine in red blood cells and the liver (Eyer & Lierheimer, 1980; Eyer et al., 1980). Whereas phenylhydroxylamine is reduced further to aniline in the liver, it can reform nitrosobenzene in red blood cells (Eyer & Lierheimer, 1980). Both nitrosobenzene and phenylhydroxylamine can produce methaemoglobin if injected *in vivo* (Kiese, 1966). Continued nitrobenzene exposure produces even more nitrosobenzene, which, when reduced to phenylhydroxylamine by NAD(P)H, can reform nitrosobenzene, thus completing the cycle (Figure 5); this cycle is referred to as a redox couple. If exposure to nitrobenzene is sufficient, nitrosobenzene and phenylhydroxylamine can form significant catalytic pools in erythrocytes, and the redox cycling expends cofactor NAD(P)H and native oxygenated haemoglobin faster than they can be regenerated. Because red blood cells have only a limited capacity to reduce methaemoglobin back to haemoglobin by methaemoglobin reductase, methaemoglobin can accumulate. This redox couple tends to resist the expedient detoxification of nitrobenzene and its intermediates.

Although Kiese (1966) and Reddy et al. (1976) showed that nitrobenzene exposure is linked to methaemoglobin formation, Goldstein & Rickert (1985) later showed that nitrobenzene does not increase methaemoglobin formation when incubated *in vitro* with Fischer-344 male rat red blood cell suspensions. This lack of activity *in vitro* was surprising, contradicting the *in vivo* results. It was not due to a lack of nitrobenzene transfer across red blood cell membranes, because [^{14}C]nitrobenzene accumulated to the same extent as labelled *o*- and *p*-dinitrobenzenes, compounds that cause methaemoglobin formation after being metabolized by glutathione transferase (Rickert, 1987). Furthermore, because red blood cell uptake of nitrobenzene was unaffected by temperature and was quickly maximal, Goldstein & Rickert (1985) reasoned that nitrobenzene uptake must be by simple partitioning, not by active transport. Lastly, if the red blood cell membrane was limiting, then red blood cell lysates should provide direct nitrobenzene access to oxyhaemoglobin, yet no methaemoglobin formed when nitrobenzene was incubated with red blood cell haemolysates. Taken together, these results are best explained by the fact that red blood cells do not have sufficient amounts of microsomal P-450 and lack mitochondria; both are likely to be necessary to reduce significant amounts of nitrobenzene. Goldstein & Rickert (1985) concluded that activated nitrobenzene intermediates are made external to red blood cells, passively transfer across the red blood cell membrane and then set up a redox cycle as outlined and interact with oxyhaemoglobin to form methaemoglobin.

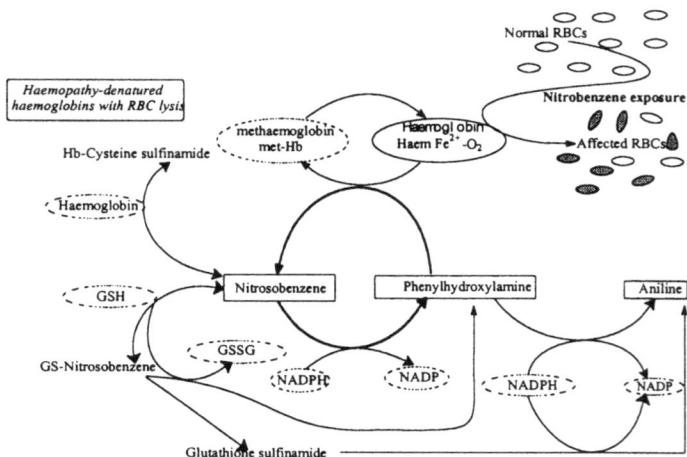

Fig. 5. Red blood cell (RBC) cycling of nitrosobenzene and phenylhydroxylamine (Holder, 1999a). The conversion of nitrosobenzene to phenylhydroxylamine is driven by the oxidation of NAD(P)H and haemoglobin-Fe^{2+}. Nitrosobenzene can outcompete for oxygen on functional tetrameric haemoglobin. Not shown is the destabilization of tetrameric haemoglobin to two haemoglobin dimers. Nitrosobenzene can also bind to cysteine groups on functional haemoglobin, thereby denaturing the globin chain. Functional oxyhaemoglobin is altered, and haemopathy occurs. Because of the redox disturbances, many red blood cells are turned over more rapidly in the spleen, leading to engorgement. Glutathione can bind nitrosobenzene, and the conjugate can move systemically and regenerate nitrosobenzene elsewhere, to again start up this pernicious cycle. Other cell types are likely to regenerate nitrosobenzene and phenylhydroxylamine, although these mechanisms are less well understood. See text for definitions.

Although red blood cells lack the microsomal and mitochondrial reductive capability of liver, they can maintain stable, steady-state levels of nitrosobenzene and phenylhydroxylamine. Nitrosobenzene can bind the haemoglobin haem-Fe^{2+} better than its normal ligand oxygen by approximately 14-fold. Nitrosobenzene also promotes the oligomeric dissociation of native oxygen-carrying haemoglobin,[1] $Hb_4 \leftrightarrow 2Hb_2$ (Eyer & Ascherl, 1987). This reaction in part explains the cyanotic presentation of patients exposed to nitrobenzene by a number of routes (e.g., Stevens, 1928; Zeligs, 1929; Zeitoun, 1959; Abbinante & Pasqualato, 1997). Although nitrosobenzene participation in haemoglobin oxidation is still not fully understood, it is known that methaemoglobin is formed when nitrosobenzene dissociates from haemoglobin (Eyer & Ascherl, 1987). Furthermore, nitrosobenzene can react with globin cysteine molecules, causing protein denaturation (Kiese, 1974). After sufficient nitrosobenzene and phenylhydroxylamine recycling, red blood cell membranes can destabilize, resulting in red blood cell lysis. This is clinically referred to as "chocolate blood plasma" and is associated with haemosiderosis and toxic anaemia (Kiese, 1966; Smith et al., 1967; Goldstein & Rickert, 1985). Cellular deposits result from erythrocyte destruction; erythrocyte debris is apparent in the spleen, but other organs are also affected.

Using ESR spin trapping techniques, Maples et al. (1990) studied the *in vitro* and *in vivo* formation of 5,5-dimethyl-1-pyrroline-*N*-oxide (DMPO)/haemoglobin thiyl or DMPO/glutathiyl free radical adducts in red blood cells and blood of male Sprague-Dawley rats and humans. They investigated whether free radicals were actually involved in red blood cells. Spin trap results indicated that aniline, phenylhydroxylamine, nitrosobenzene and nitrobenzene can all be metabolized *in vivo* to yield a common metabolite, which could trapped by DMPO. Their ESR data show that the common metabolite is the phenylhydronitroxide radical. Maples et al. (1990) postulated that free radicals within red blood cells are responsible for the oxidation of thiol groups. Nitrosobenzene reversibly reacts with glutathione (GSH) to form the GS–nitrosobenzene conjugate. Glutathione conjugates can be transported systemically and, under some conditions, can deconjugate at distal organ sites (Klaassen, 2001). GS–nitrosobenzene can react as follows: (1) with haemoglobin oxygen to form sulfhaemoglobin; (2) with another molecule of GSH to form GSSG (oxidized dimer) and phenylhydroxylamine (Eyer, 1979); or (3) by rearrangement to form

---

[1] This denaturation reaction is shown in the extreme here. Variations are likely to occur with fewer than four oxygen molecules removed; i.e., three, two or one may be removed, leaving various levels of haemoglobin oxygen saturation. A mixture of reacted haemoglobin may be expected, depending on the blood concentration of nitrobenzene.

glutathione sulfinamide (GSO-aniline) and then be reduced to aniline (Eyer & Lierheimer, 1980). The latter two reactions are shown in Figure 5. The third reaction sequence is thought to be a major pathway for conversion of nitrosobenzene to aniline (Eyer & Lierheimer, 1980). Extensive nitrobenzene exposure, leading to correspondingly extensive glutathione conjugation, can deplete cellular glutathione supplies over time.

Phenylhydroxylamine can react with oxyhaemoglobin to form the reactive phenylhydronitroxide free radical, which can then react with glutathione to form the thiyl free radical (GS·), a radical found *in vivo* and *in vitro* (Maples et al., 1990). The glutathiyl radical can pass its unpaired electron to various natural trapping molecules (e.g., ascorbate, α-tocopherol), certain unsaturated fatty acids and other cellular components (Keher, 1993; Gutteridge, 1995). The range of activities depends on the amounts and stabilities of the free radical intermediates.

Catalase is a catabolic enzyme that catalyses the rapid decomposition of hydrogen peroxide in red blood cells; following the absorption of nitrobenzene, catalase activity is inhibited (Goldstein & Popovici, 1960). Catalase has been reported to be inhibited by very low concentrations of the nitrobenzene metabolites phenylhydroxylamine and *p*-aminophenol (De Bruin, 1976). With respect to scavenging of peroxides in biological systems, there are several glutathione peroxidases that can scavenge hydrogen peroxide, as well as certain organic peroxides. In an *in vitro* test system, 20–200 times the amount of hydrogen peroxide was necessary to produce methaemoglobin in normal red blood cell suspensions (i.e., with catalase) than in red blood cells in which catalase was lacking; inhibition of catalase in red blood cells by nitrobenzene may contribute to its potency as a methaemoglobin producer (De Bruin, 1976).

6.4 Elimination and excretion

6.4.1 Oral exposure

The major route of excretion after oral exposure to nitrobenzene is the urine. After a single oral administration via gavage of 25 mg [^{14}C]nitrobenzene/kg of body weight to rats, 50% of the nitrobenzene dose appeared in the urine within 24 h, and 65% after 7 days.

Excretion in the faeces within 7 days was 15.5%. Thus, about 80% of the dose could be accounted for in the excreta in this study (Albrecht & Neumann, 1985). Excretion via urine and faeces was ostensibly complete after 3 days.

In rats (Fischer-344 and Sprague-Dawley strains) given single oral doses of 22.5 or 225 mg [^{14}C]nitrobenzene/kg of body weight, approximately 72–88% of the administered dose was recovered in 72 h, of which about 80% was in urine (Rickert et al., 1983). (Equivalent results to those after oral dosing were obtained after intraperitoneal dosing of Fischer-344 rats with 225 mg/kg of body weight.) For the metabolite pattern, see section 6.3.1.

In rabbits given a single dose of [^{14}C]nitrobenzene by gavage, 70% of the radioactivity was eliminated from the animals in expired air, urine and faeces within 4–5 days after dosing. The remainder of the radioactivity was found to be slowly excreted in urine and possibly in expired air (as carbon dioxide). Details of the metabolic profile of excreted radioactivity are given in section 6.3.1.

In cases of human poisoning, metabolites identified in the urine have been *p*-aminophenol and *p*-nitrophenol (Von Oettingen, 1941; Ikeda & Kita, 1964; Myślak et al., 1971).

Following the oral intake of a single dose of 30 mg nitrobenzene in one volunteer, urinary excretion of *p*-nitrophenol occurred slowly. The initial half-time of elimination of *p*-nitrophenol was around 5 h, with a late-phase half-time of >20 h (estimated from a figure). Since the intake of a 5-mg oral dose of *p*-nitrophenol by the same subject led to the very rapid urinary excretion of *p*-nitrophenol (all eliminated by 8 h), it appears that the slow excretion of *p*-nitrophenol after dosing with nitrobenzene is due to inhibition of metabolism at the higher dose (Piotrowski, 1967).

An examination of human poisoning cases (section 8.1.1) did not provide any useful information about the rate of nitrobenzene elimination; reports of accidental or intentional ingestion indicated that recovery took from "several days" (after ingestion of about 7 ml) to >40 days (after ingestion of about 40 ml).

6.4.2   Dermal exposure

Following exposure to nitrobenzene vapours (without inhalation exposure), a proportion of nitrobenzene (quantity not reported) absorbed through the skin was excreted in the urine of volunteers as *p*-nitrophenol (Piotrowski, 1967).

6.4.3   Inhalation exposure

Urinary excretion of *p*-nitrophenol was found in seven volunteers who had inhaled 5–30 mg nitrobenzene/m³ for 6 h (Salmowa et al., 1963). The rate of urinary elimination showed considerable interindividual variation but was broadly dose dependent. In general, excretion was most rapid during the first 2 h and then levelled off. The elimination of *p*-nitrophenol in urine had estimated (from a figure) half-lives of about 5 and >70 h. In some cases, *p*-nitrophenol could be detected for as long as 100 h after exposure. In a 47-year-old woman who had been occupationally exposed to nitrobenzene for 17 months, sufficient to cause symptoms of toxicity, *p*-nitrophenol and *p*-aminophenol were found in the urine, gradually being eliminated over 2 weeks; levels of *p*-nitrophenol were between 1 and 2 times higher than levels of *p*-aminophenol (Ikeda & Kita, 1964). In four men exposed to nitrobenzene at 10 mg/m³ by the inhalational route (using a system designed to exclude dermal absorption of nitrobenzene vapour) for 6 h per day over a number of days (absorbed amounts ranged between 18.2 and 24.7 mg per daily exposure), a mean value of 16% of the absorbed dose was excreted in the urine as *p*-nitrophenol (Piotrowski, 1967), a value in close agreement with an earlier value of 13%, obtained after single exposures (Salmowa et al., 1963). The half-time of elimination of *p*-nitrophenol was not estimated, but excretion was followed for 3 days after exposure ceased.

An examination of several literature cases of poisoning of children from overnight inhalational exposure to nitrobenzene (section 8.1.3) did not provide any useful information about the rate of elimination of nitrobenzene, noting only that recovery from clinical signs took from 4 to 8 days.

6.5   Retention and turnover

6.5.1   *Protein binding* in vitro

The binding of [^{14}C]nitrobenzene to plasma of rainbow trout (*Oncorhynchus mykiss*) and Sprague-Dawley rats was determined *in vitro* using a centrifugal microfiltration system to separate bound from free label, following equilibrium binding (Schmieder & Henry, 1988). Nitrobenzene was 79.4% bound to trout plasma and 72% bound to rat plasma.

6.5.2   *Body burden and (critical) organ burden*

A study by Freitag et al. (1982) indicated that, after administration of three daily oral doses of nitrobenzene to rats, there was no evidence of significant retention of nitrobenzene or its metabolites in the body (see also section 6.2.1); several other toxicokinetic studies (see section 6.4) suggest that after single oral doses in rats, between about 81% and 88% of the dose was recovered in urine and faeces.

6.6   Reaction with body components

The covalent binding of [^{14}C]nitrobenzene was investigated in erythrocytes and spleens of male B6C3F$_1$ mice and male Fischer-344 rats following single oral doses (Goldstein & Rickert, 1984). Total and covalently bound ^{14}C concentrations in erythrocytes were 6–13 times greater in rats than in mice following administration at doses of 75, 150, 200 and 300 mg/kg of body weight. Covalently bound ^{14}C in erythrocytes peaked at 24 h in rats after 200 mg/kg of body weight, whereas the low level of binding plateaued in mice at 10 h. Gel filtration and polyacrylamide gel electrophoresis revealed that haemoglobin was the primary, if not exclusive, site of macromolecular covalent binding. Splenic engorgement increased in a time-related manner after nitrobenzene dosing in rats but not in mice; covalent binding of nitrobenzene and its metabolites in spleen was primarily derived from bound ^{14}C from scavenged erythrocytes. The peak of splenic binding of ^{14}C in rats occurred between 24 and 48 h, whereas levels of splenic binding in mice were very low, being about one-fifteenth of those in rats. It appears that the degree of erythrocytic damage in mice is not sufficient to elicit splenic scavenging and clearance from the systemic circulation. Thus, the species difference in splenic engorgement and splenic accumulation of nitrobenzene is likely to be related to differences in susceptibility to nitrobenzene-induced red blood cell damage, which in turn may be related to the species difference in distribution of nitrobenzene or its metabolites to the erythrocytes.

To establish haemoglobin adduct formation as a possible means of biological exposure monitoring, Albrecht & Neumann (1985) used GC methods to determine aniline in the hydrolysates of haemoglobin from rats orally dosed 24 h previously with unlabelled aniline hydrochloride or nitrobenzene. The binding indices for unlabelled nitrobenzene were very similar to those obtained in experiments using labelled nitrobenzene, with the binding index after aniline dosing being about one-quarter that after nitrobenzene dosing. Results suggested that 4–5 times as much nitrosobenzene is formed from nitrobenzene as from an equal amount of aniline (see section 6.7).

The haemoglobin binding of five nitroarenes — i.e., nitrobenzene, 4-nitrobiphenyl, 1-nitropyrene, 2-nitronaphthalene and 2-nitrofluorene — and their corresponding amines, administered orally to male SD rats, was determined by HPLC to evaluate the extent of *in vivo* reductive and oxidative activations of these compounds to N-hydroxylamines, which covalently bind to haemoglobin to form acid-labile sulfinamides (Suzuki et al., 1989). Except for nitrobenzene, haemoglobin binding of the nitroarenes was significantly lower than that of the corresponding amines. Haemoglobin binding of nitrobenzene and 4-nitrobiphenyl decreased markedly after pretreatment with antibiotics, indicating that the reductive activation of nitrobenzene and 4-nitrobiphenyl is largely dependent upon metabolism by intestinal microflora; the binding of the other compounds and of aniline did not decrease appreciably (Suzuki et al., 1989).

A series of 21 nitroarenes, including nitrobenzene, was given to female Wistar rats by gavage (0.1 ml of 0.5 mol/litre solutions per 100 g of body weight), and blood samples were taken by heart puncture 24 h later. Hydrolysable haemoglobin adducts were determined by GC/MS (Sabbioni, 1994). Nitrobenzene formed adducts with haemoglobin, findings in agreement with those of Suzuki et al. (1989). It was concluded that, except for a few outliers, the extent of haemoglobin binding increases with the reducibility of the nitro group.

Haemoglobin, either in the intact red blood cells or in haemo-lysates, readily reacts with mono- and dinitrobenzoates (Norambuena et al., 1994). The measured reactivity in oxidizing haemoglobin in *in vitro* erythrocyte suspensions followed the order *m*-nitrobenzoic acid > 3,5-dinitrobenzoic acid > *p*-nitrobenzoic acid > *o*-nitrobenzoic acid >> nitrobenzene. The rate of the process was faster in haemolysates than in whole red blood cells. At low concentrations of nitroaromatics (<8 mmol/litre), almost quantitative production of methaemoglobin was observed; at higher concentrations, however, the kinetics became complex, and other haemoglobin derivatives were produced (Noram-buena et al., 1994).

6.7   Biomarkers of exposure

The presence of methaemoglobinaemia can indicate exposure to nitrobenzene. However, this condition in itself cannot be used as a specific biomarker of exposure to nitrobenzene, since other toxic sub-stances can also have the same effect. Similarly, other measures of toxic damage to haemoglobin can be used, including the formation of haemichromes and Heinz bodies, as well as erythrocytic levels of reduced glutathione (Luke & Betton, 1987); as noted, however, such end-points are common to other drugs and chemicals that oxidize the haem iron of oxyhaemoglobin to form methaemoglobin.

The presence of *p*-aminophenol in the urine can be detected at high levels of exposure, but it is undetectable at low levels of exposure that are considered to be "safe" (Ikeda & Kita, 1964; Piotrowski, 1967). Piotrowski (1967) suggested that urinary *p*-nitrophenol can be used as a test of nitrobenzene exposure that appears to be highly specific in practice; if the daily exposure is essentially constant (e.g., in an occupational setting), a good estimate of the mean daily intake can be calculated from data obtained by taking urine specimens on each of the last 3 days of the working week. However, it should be noted that this compound is also a biomarker of exposure to the insecticide parathion (Denga et al., 1995). Estimates of nitrobenzene intake following single exposures have also been made (Salmowa et al., 1963). Harmer et al. (1989) measured *p*-nitrophenol in the urine of workers exposed to nitrobenzene in the air, suggesting that it was a suitable index of exposure to nitrobenzene; aminophenols that can occur at higher levels of nitrobenzene exposure can be found in urine from physiological and other sources.

The nitrobenzene metabolites nitrosobenzene and hydroxylaminobenzene have been found to produce methaemoglobinaemia, and ^{14}C from labelled nitrobenzene (or its metabolites) is covalently bound to haemoglobin in the blood of orally exposed mice and rats (Goldstein et al., 1983a; Goldstein & Rickert, 1984); the presence of these haemoglobin adducts may serve as a biomarker of exposure, although they are difficult to quantify, and assays are expensive. Albrecht & Neumann (1985) concluded that the determination of aniline by GC after hydrolysis of haemoglobin adducts could be developed as a means of biological exposure monitoring for nitrobenzene; they were able to determine the aniline cleavage product in the nanogram range (10 pg absolute), suggesting that the method has high sensitivity. A further advantage of this measure over measurement of methaemoglobin was the relative stability of the haemoglobin adducts over time. Thus, the method may provide a means to estimate cumulative exposure over weeks or some months. One problem may be that human haemoglobin may form adducts with nitrosobenzene less extensively than rat haemoglobin; in an *in vitro* study in which rat and human erythrocytes were incubated with 0.37 mmol nitrosobenzene/litre, rat haemoglobin contained 3–4 times as much adduct as the human haemoglobin. Measurement of haemoglobin adducts is also thought to predict the macromolecular damage that leads to critical toxicity in other potential target tissues (Neumann, 1988). As the metabolism that predisposes the binding may vary individually, the method has the potential to take account of individual susceptibility.

7. EFFECTS ON LABORATORY MAMMALS AND *IN VITRO* TEST SYSTEMS

7.1 Single exposure

7.1.1 Oral

The oral administration of 4 ml (4.8 g) of nitrobenzene was reported to be almost instantly fatal to young rabbits, whereas cats died after 24 h following administration of 2 ml (2.4 g) of nitrobenzene. The minimal fatal dose in dogs was stated to be 750–1000 mg/kg of body weight by the oral route (Von Oettingen, 1941).

Sziza & Magos (1959) reported an oral LD_{50} in female white rats (170–200 g of body weight; strain not stated) of 640 mg/kg of body weight (administered in 10% gum arabic in water); at this dose, the percentage of methaemoglobin in the blood was 11%, 19% and 28% at 0.5, 1 and 2 h after dosing, respectively. The LD_{50} for rats has been reported as 600 mg/kg of body weight (Smyth et al., 1969).

In female giant chinchilla rabbits given gavage doses of nitrobenzene in water, doses leading to death ranged between 180 and 370 mg/kg of body weight; above 200 mg/kg of body weight, at least 50% of the animals died. The toxicity was not immediately apparent. At doses of the order of 250 mg/kg of body weight, death occurred between 2 and 6 days after dosing. At doses above 300 mg/kg of body weight, death occurred within 24 h (Robinson et al., 1951).

Morgan et al. (1985) administered a single oral dose of nitrobenzene at 550 mg/kg of body weight to male Fischer-344 rats. Within 24 h, rats were lethargic and ataxic; within 36–48 h, they displayed moderate to severe ataxia and loss of righting reflex and no longer responded to external stimuli. By 48 h, petechial haemorrhages were observed in the brain stem and cerebellum and bilaterally symmetric degeneration (malacia) in the cerebellum and cerebellar peduncles. Tracer studies indicated that a very small percentage of nitrobenzene reached the brain, and it was present as parent compound. It accumulated at a higher concentration in grey than in white matter, but there was no preferential accumulation in areas where lesions occurred.

Bond et al. (1981) exposed Fischer-344 rats (six males per group) to single oral doses of nitrobenzene of 50, 75, 110, 165, 200, 300 or 450 mg/kg of body weight; three rats per group were sacrificed 2 and 5 days later. Histopathological changes consistently involved the liver and testes. Centrilobular hepatocytic necrosis appeared in rats 2 or 5 days after administration of ≥200 mg/kg of body weight, whereas hepatocellular nucleolar enlargement was detected at ≥110 mg/kg of body weight. Necrosis of spermatogenic cells as well as multi-nucleated giant cells were seen between 1 and 4 days after administration of 300 mg/kg of body weight or more. Necrotic debris and decreased numbers of spermatozoa in the epididymis were observed after 5 days at 300 mg/kg of body weight and between 2 and 5 days at 450 mg/kg of body weight. One high-dose rat had a microscopic cerebellar lesion. The above findings were not ascribed by the authors to methaemoglobinaemia, since sodium nitrite, which produces an equivalent haematological effect, did not produce any of these histo-pathological changes.

7.1.2 Dermal

Shimkin (1939) reported that the minimal fatal dermal dose in mice was about 0.4 ml/kg of body weight (480 mg/kg of body weight). Neat nitrobenzene was painted onto the shaved abdomens of female C3H mice over less than 10% of the body surface area. Fifteen of 18 collapsed within 1 h, but all recovered by 24 h. A repeat application led to the death of 3 of 18 on day 3, whereas a third application led to the death of a further 9 within 48 h. Ten strain-A mice were painted "vigorously" over the unshaved abdomen for about 20 s. All were in partial collapse within 30 min, and 8 of 10 died between 3 h and 3 days. After high dermal or subcutaneous doses, the mice were prostrate within 30 min and lay either motionless or with occasional twitching movements. Some recovered, while others died. By 3 h, the white blood cell count had dropped significantly from 11 000–14 000/mm^3 to 5000/mm^3 (mm^3 = µl), with a normal differential count; the red blood cell count was not affected. At 21 h, the white blood cell count was 1000–1500/mm^3 or lower, with some reduction in red blood cell count and evidence of hypochromia and haemolysis. Necropsy findings included chocolate-coloured blood, dark grey-blue skin, orange urine with a nitrobenzene odour and livers that were white and soft; other organs (not stated whether this included the brain) were grossly normal. Outer portions of the liver lobules showed diffuse necrosis; hepatocyte cell cytoplasm was pale and granular, and many nuclei were not visible. Kupffer cells had large amounts of dark brownish pigment. Kidneys showed slight swelling of the glomeruli and tubular epithelium. The spleen, lung and testis were morphologically normal.

After administration of the high dermal doses sufficient to prostrate the female C3H mice within 30 min, the blood became chocolate coloured and more viscous at 1–3 h after exposure, and the skin developed a dark grey-blue hue; spectral analysis of the blood revealed methaemoglobin (Shimkin, 1939).

In female white rats (170–210 g of body weight; strain not stated), Sziza & Magos (1959) determined the acute dermal toxicity of nitrobenzene (LD_{50}) to be 2100 mg/kg of body weight. At the dermal LD_{50}, the percentage of methaemoglobin in the blood was 16%, 25% and 35% at 0.5, 1 and 2 h, respectively. At gross necropsy, organs were hyperaemic, and the liver was strongly reddish-brown in colour. Histopathology revealed hyperaemia and degeneration of the parenchymal organs, and marked fine droplet fatty degeneration of the liver and kidneys was noted. For comparison, the acute oral LD_{50} in these experiments was 640 mg/kg of body weight.

Nitrobenzene at doses of 560, 760 or 800 mg/kg of body weight was applied to the clipped trunks of New Zealand albino rabbits (five per group) for 24 h. Methaemoglobinaemia was seen in less than 20 min at all doses. At the low dose, there were no mortalities, but persistent lethargy and discoloration of skin and eyes were apparent. At the middle and high doses, 80% mortality was observed within 48–96 h and 24–48 h, respectively. The LD_{50} was approximated at 760 mg/kg of body weight (Harton & Rawl, 1976).

7.1.3   *Inhalation*

In an inhalation study, groups of 12 male or female Sprague-Dawley rats were exposed to an atmosphere saturated with nitrobenzene vapour for 3 or 7 h. None of the 12 rats died during 3 h of exposure or within the postexposure period of 14 days, but 3 of 12 rats died after 7 h of exposure (BASF, 1977).

7.1.4   Intraperitoneal

In female white rats (170–210 g of body weight; strain not stated), Sziza & Magos (1959) determined the acute intraperitoneal toxicity of nitrobenzene (LD_{50}) to be 640 mg/kg of body weight, equivalent to the acute oral LD_{50}. Histopathology revealed hyperaemia and degeneration of the parenchymal organs and marked fine droplet fatty degeneration of the liver and kidneys. At a dose of 640 mg/kg of body weight, equivalent to the LD_{50}, the percentage of methaemoglobin in the blood was 30.5%, 39% and 33.5% at 0.5, 1 and 2 h, respectively.

7.2   Short-term exposure

Most repeated-dose studies have concentrated on methaemo-globinaemia and its consequences on blood elements, spleen and liver. These findings are described in section 7.7. Findings on reproductive function are described in section 7.5.

7.2.1   Oral

A 28-day repeated-dose gavage study with nitrobenzene was performed in male and female Fischer-344 rats (six per sex per group) at doses of 0, 5, 25 and 125 mg/kg of body weight per day (Shimo et al., 1994). An additional two groups of animals exposed to 0 or 125 mg/kg of body weight per day were kept for a 2-week recovery period. One female in the 125 mg/kg of body weight per day group died on day 27. Decreased movement, pale skin, gait abnormalities and decreases in body weights or body weight gains were seen at the high dose. Increases of total cholesterol and albumin and decreases in blood urea nitrogen were seen at the middle and high doses, and increases of albumin/globulin ratio in both sexes and alanine amino-transferase, alkaline phosphatase and total protein were observed in females in the high-dose group. Increases in weights of the liver, spleen and kidney and decreases in the weights of the testis and thymus were seen at the high dose. Increased liver weights were also seen in low-dose males, whereas increased spleen weights were seen in both sexes at the middle dose. Histopathology revealed spongiotic changes and brown pigmentation in the perivascular region of the cerebellum, degeneration of seminiferous tubular epithelium and atrophy of seminiferous tubules at 125 mg/kg of body weight per day. The above findings disappeared or tended to decrease during or at the end of the recovery period. No no-observed-effect level (NOEL) was established in this study (Shimo et al., 1994).

In a range-finding US National Toxicology Program (NTP) study, nitrobenzene was administered to $B6C3F_1$ mice and Fischer-344 rats (both sexes) by gavage at doses in the range 37.5–600 mg/kg of body weight per day for 14 days (NTP, 1983a). All rats and mice at the high dose of 600 mg/kg of body weight per day and all rats at 300 mg/kg of body weight per day died or were sacrificed in a moribund condition prior to the end of treatment. Treated animals were inactive, ataxic, prostrate, cyanotic and dyspnoeic. Significant depression of weight gain (>10%) was seen in male mice at 37.5 mg/kg of body weight per day and in mice of both sexes at 75 mg/kg of body weight per day. Histologically, mice and rats showed changes in the brain, liver, lung, kidney and spleen.

In an NTP study, nitrobenzene was administered to $B6C3F_1$ mice (10 per sex per group) by gavage at doses of 0, 18.75, 37.5, 75, 150 or 300 mg/kg of body weight per day for 13 weeks (NTP, 1983a). Mean final body weights were not affected. Three high-dose males died or were sacrificed moribund in weeks 4 and 5. Clinical signs included ataxia, lethargy, dyspnoea, convulsions, irritability and rapid head-bobbing movements. Liver weight in treated mice was increased compared with controls; the increase was statistically significant at the two highest doses in males and at all doses in females. Fatty change was reported in the X-zone (basophilic cells that surround the medulla around 10 days of age, then gradually disappear as mice mature) of the adrenal glands of 8 of 10 high-dose female mice. One high-dose male had acute necrosis in the area of the vestibular nucleus in the brain. No NOEL could be derived from this study.

Nitrobenzene was administered to Fischer-344 rats (10 per sex per group) by gavage (in corn oil) at doses of 0, 9.375, 18.75, 37.5, 75 or 150 mg/kg of body weight per day for 13 weeks (NTP, 1983a). Mean final body weights were not affected. Seven high-dose male rats died, and 2 of 10 were sacrificed moribund during weeks 10, 11 and 13. One high-dose female died and two were sacrificed during weeks 6, 7 and 9. Clinical signs included ataxia, left head tilt, lethargy, trembling, circling and dyspnoea, as well as cyanosis of the extremities in the two highest dose groups in both sexes.

Brain lesions were found in 8 of 10 males and 7 of 10 females at 150 mg/kg of body weight per day; the lesions appeared to be localized in the brain stem to areas of the facial, olivary and vestibular nuclei and to cerebellar nuclei and probably correlate with the clinical findings of head tilt, ataxia, trembling and circling. These lesions were characterized by demyelination, loss of neurons, varying degrees of gliosis, haemorrhage, occasional neutrophil infiltration and, occasionally, the presence of haemosiderin-containing macrophages. Brain vascular lesions (as described in the rat dermal study; see section 7.2.2 below) were not observed in this gavage study.

In an immunotoxicology study (see section 7.8), Burns et al. (1994) dosed female $B6C3F_1$ mice with 0, 30, 100 or 300 mg nitrobenzene/kg of body weight per day in corn oil by gavage for 14 days, with necropsy on day 15; the high dose was close to a maximum tolerated dose (MTD), with 8.5% of animals dying during the exposure period. There was a slight dose-related increase in body weight that was statistically significant at the high dose. Liver and spleen appeared to be the primary target organs for toxicity, with dose-dependent increases in weight (significant only at the two highest doses). Dose-related increases in alanine aminotransferase and aspartate aminotransferase, marginal at the low dose, were suggestive of liver toxicity, whereas gross histopathology revealed mild hydropic degeneration around focal central veins in the liver (high dose only).

Using the OECD Combined Repeat Dose and Reproductive/ Developmental Toxicity Screening (ReproTox) test protocol, nitrobenzene was given by gavage to Sprague-Dawley rats (10 per sex per group) at 0, 20, 60 or 100 mg/kg of body weight throughout premating (14 days), mating (14 days), gestation (22 days) and lactation (4 days); females and pups were necropsied at this stage, while surviving males were killed at day 41 or 42 (Mitsumori et al., 1994). At 100 mg/kg of body weight, animals exhibited piloerection, salivation, emaciation and anaemia from day 13. Additionally, some animals exhibited neurological signs, with deaths of two males and nine females. High-dose animals showed reduced food consumption a week or so after dosing, and body weight gain was significantly depressed during the study. Six mid-dose females showed anaemia, with neurological signs in one and, during the lactation period, reduced food consumption and body weight gain. Blood biochemical changes indicative of liver toxicity were reported. Increased absolute and relative organ weights of liver and spleen were seen in treated males; smaller increases were seen in the absolute and relative kidney weights of mid- and high-dose males. Neuronal necrosis and gliosis were observed in certain nuclei in the cerebellar medulla and pons in mid- and high-dose males (respective incidence of 3/10 and 10/10 compared with 0/10 for controls and low-dose rats).

7.2.2 *Dermal*

Dermal applications of nitrobenzene by skin painting (dose not stated) to female C3H or male strain-A mice resulted in methaemoglobinaemia by 3 h after application and the death of 12 of 18 and 8 of 10 animals, respectively. Although two or three applications were required for the C3H mice, most animals were in partial collapse within 15 min and dead by the third day. Most of the strain-A mice were dead within the first day. The liver was reported to be the most severely affected organ, with a diffuse necrosis in the outer two-thirds of the lobules. Histopathological examination of the kidneys noted a slight swelling of the glomeruli and tubular epithelium (Shimkin, 1939).

A study by Matsumaru & Yoshida (1959), mainly aimed at investigating the possible synergism between alcohol intake and nitrobenzene poisoning, included a group of five rabbits (strain not stated) that were dermally treated with nitrobenzene (dose not stated; possibly 0.1 ml per day) at intervals of 3 days (for 57–63 days), 7 days (for 22 days) or 27–32 days (for 116 days). Reported findings included well defined, round vacuoles in the medulla of the brain, congestion, oedema, emphysema and atelectasis in the lungs, some congestion and swelling in the liver (with several animals showing hepatocellular degeneration) and kidney congestion.

In a range-finding NTP study, nitrobenzene was administered to B6C3F$_1$ mice and Fischer-344 rats (both sexes) by skin painting at doses in the range 200–3200 mg/kg of body weight per day for 14 days (NTP, 1983b). All rats and mice at the 1600 and 3200 mg/kg of body weight per day doses died or were sacrificed moribund prior to the end of treatment. Treated animals were inactive, ataxic, prostrate and dyspnoeic. Significant depression of weight gain (>10%) was seen in mice from all dose groups. Histologically, mice and rats showed changes in the brain, liver, spleen and testes, with mice less affected than rats.

In an NTP study, nitrobenzene was administered to B6C3F$_1$ mice (10 per sex per group) by skin painting (in acetone vehicle) at 0, 50, 100, 200, 400 or 800 mg/kg of body weight per day for 13 weeks (NTP, 1983b); the chemical was applied to a shaved area of the skin in the intercapsular region. Mean final body weights were not significantly affected. Six high-dose males were sacrificed moribund, and three died between weeks 3 and 10; seven high-dose females were sacrificed moribund, and one high-dose female and one female of the 100 mg/kg of body weight per day group died between weeks 2 and 9. Clinical signs in some animals at the high dose included inactivity, leaning to one side, circling, dyspnoea, prostration and, in one, head tilt, whereas a number of dosed females had extremities cold to the touch. One high-dose female exhibited tremors, and two were insensitive to painful stimuli. Inflammation of the skin (diffuse or focal and of minimal to mild severity) was seen at the site of nitrobenzene application at the two highest doses; inflammatory cells were present in the dermis, with varying degrees of involvement of the subcutaneous tissue. There was acanthosis and hyperkeratosis of the epidermis, with occasional thick crusts of necrotic cells or focal areas of necrosis extending deep into the epidermis. Liver weights in treated male mice from the 400 mg/kg of body weight per day group and females from the 400 and 800 mg/kg of body weight per day groups were significantly increased compared with controls. At the high dose, a number of periportal hepatocytes were smaller than those in control livers and in treated mice, and there was a noticeable variation in the size of hepatocyte nuclei, especially in the centrilobular zone. The cytoplasm of hepatocytes in many treated mice had a homogeneous eosinophilic appearance, whereas that in controls had a vacuolated appearance characteristic of glycogen-containing cells. While degeneration of the "X" zone of the adrenal glands (the zone of cells adjacent to the medulla) in female mice was noted, the degree of vacuolation in treated animals was reported to be greater than normally seen in controls. Brain lesions were found in 2 of 10 males and 3 of 10 females at 800 mg/kg of body weight per day; the lesions appeared to be localized in the brain stem in the area of the vestibular nucleus and/or cerebellar nuclei; one high-dose female had a mild bilateral lesion in a nucleus of the ventrolateral thalamus. Such lesions were probably responsible for the clinical behavioural findings of head tilt, leaning to one side and circling. Brain vascular lesions (as described in the rat dermal study; see below) were not observed in this mouse dermal study.

No clear NOEL was established in this study, with the following findings (among others) noted at the lowest dose of 50 mg/kg of body weight per day: lung congestion, adrenal cortical fatty change and variation in the size of hepatic nuclei, especially the centrilobular zone.

Nitrobenzene was administered to Fischer-344 rats (10 per sex per group) by skin painting (in acetone vehicle) at 0, 50, 100, 200, 400 or 800 mg/kg of body weight per day for 13 weeks (NTP, 1983b); the chemical was applied to a shaved area of the skin in the intercapsular region. Mean final body weights were not significantly affected; the body weights in the high-dose group were not analysed due to a high incidence of early deaths. Seven high-dose male rats died and 3 of 10 were sacrificed moribund between weeks 4 and 10; five high-dose females died and five were sacrificed between weeks 2 and 12. Clinical signs in high-dose males included ataxia, head tilt, lethargy, trembling, circling, dyspnoea, forelimb paresis, splayed hindlimbs, diminished pain response and reduced righting response. Except for dyspnoea in a few females, the other clinical signs were not noted in females. The extremities of a number of rats (both sexes) were cold to the touch and/or cyanotic. Brain lesions were found in both sexes at 800 mg/kg of body weight per day; the lesions appeared to be localized in the brain stem to areas of the facial, olivary and vestibular nuclei and to cerebellar nuclei and probably correlate with the clinical behavioural findings. These lesions were characterized by demyelination, loss of neurons, varying degrees of gliosis, haemorrhage, fibrin in and around small vessels and occasional capillary proliferation. The brain vascular lesions were characterized by fibrin in and around vessel walls; red blood cells within macrophages at the site of haemorrhage indicated that the effect was real, not an agonal change or secondary to tissue mishandling at sacrifice. Perivascular haemosiderin-containing macrophages were occasionally observed. Brain vascular lesions as described in this dermal study were not observed in the Fischer-344 rat gavage study (see section 7.2.1) or in the B6C3F$_1$ mouse dermal study (see above).

No clear NOEL was established in this study, with lung congestion and fatty change in the adrenal cortex (in addition to the haematological findings described in section 7.7) noted at the lowest dose of 50 mg/kg of body weight per day.

7.2.3   *Inhalation*

In a Chemical Industry Institute of Toxicology (CIIT) study, Medinsky & Irons (1985) exposed 8- to 9-week-old Fischer-344 rats, Sprague-Dawley (CD) rats and B6C3F$_1$ mice (10 per sex per dose) to nitrobenzene at concentrations of approximately 0, 51, 180 or 640 mg/m^3 (0, 10, 35 or 125 ppm) via inhalation for 6 h per day, 5 days per week, for 2 weeks. At an exposure level of 640 mg nitrobenzene/m^3, there were severe clinical signs and a 40% rate of lethality in Sprague-Dawley rats after the fourth day and morbidity of all B6C3F$_1$ mice, necessitating their early sacrifice; surviving Sprague-Dawley rats, exhibiting rapid shallow breathing, wheezing and orange urogenital staining, were sacrificed at the end of the first week. In contrast, Fischer rats tolerated this level for 2 weeks without any adverse clinical signs. Significant concentration-dependent increases in relative liver, spleen and kidney weights were reported, primarily in Fischer rats; relative spleen weights were increased as much as 3 times those of control in Fischer rats and were still greater than controls in recovery animals ($n = 5$) at 14 days after exposure. Kidney and liver weights had recovered by day 14, but not by day 3, after exposure.

In another CIIT study (Hamm, 1984; Hamm et al., 1984), Fischer-344 rats, Sprague-Dawley (CD) rats and B6C3F$_1$ mice (10 per sex per dose per species or strain; 6–8 weeks old at the start of exposure) were exposed to nitrobenzene vapour concentrations of 0, 26, 82 and 260 mg/m^3 (0, 5, 16 and 50 ppm) for 6 h per day, 5 days per week, for 90 days. There was no effect on body weight gain or mortality. Spleen and liver weights were increased at 260 mg/m^3 in mice and rats and in the rat strains only at 82 mg/m^3. Kidney weights were increased at 260 mg/m^3 in male CD rats. Both rat strains had reduced testicular weights at 260 mg/m^3.

A variety of histopathological findings reported in the two inhalation studies are outlined below.

Liver lesions reported in the 2-week study (Medinsky & Irons, 1985) included centrilobular necrosis and severe hydropic degeneration in B6C3F$_1$ mice at 640 mg/m^3, hepatocyte necrosis in male CD rats at 180 mg/m^3 and sinusoidal congestion, centrilobular hydropic degeneration and basophilic hepatocyte degeneration in periportal areas at 640 mg/m^3 in rats that died early. Although Fischer-344 rats showed dose-related increases in relative liver weights, there were no significant histological findings. In the 90-day study, hepatocyte hyperplasia and multinucleated hepatocytes, which were more severe in males, were reported in B6C3F$_1$ mice exposed at 82 mg/m^3. CD rats had primarily a centrilobular hepatocyte hyper-trophy, with some cells containing enlarged nucleoli (82 and 260 mg/m^3), increased cytoplasmic basophilia in periportal hepato-cytes and microgranulomas (all exposure levels). Fischer-344 rats exhibited centrilobular necrosis and disorganization of hepatic cords, primarily but not exclusively in 260 mg/m^3 animals (Hamm et al., 1984).

In the 2-week study, renal effects in B6C3F$_1$ mice included mini-mal to moderate multifocal degenerative changes in tubular epithelium of males exposed to 180 mg/m^3. Neither hydropic degeneration of the cortical tubular cells nor hyaline nephrosis was seen, even at the highest exposure level of 640 mg/m^3. In CD rats, hydropic degenera-tion of the cortical tubular cells was observed (20% of males; 90% of females) at 640 mg/m^3. Renal lesions in Fischer-344 rats at 640 mg/m^3 included reversible, moderate to severe hyaline nephrosis (Medinsky & Irons, 1985). In the 13-week study, dose-related renal toxic nephro-sis was observed as the main finding in male rats of both strains (not significant at 26 mg/m^3 in CD rats), but not in mice (Hamm, 1984); the lesion was described as an accumulation of hyaline or eosinophilic droplets in the cytoplasm of proximal tubular epithelial cells.

In the 2-week study, bilateral perivascular haemorrhage in the cerebellar peduncle, accompanied by varying degrees of oedema and malacia (cell breakdown), was observed at 640 mg/m^3 in 8 of 19 mice (both sexes) sacrificed at 2–4 days and in 14 of 19 SD rats (both sexes) sacrificed in a moribund condition (Medinsky & Irons, 1985). No brain lesions were found in Fischer rats. No neurological signs were reported by Hamm et al. (1984) in either mice or rats exposed to 26, 82 or 260 mg/m^3 for 90 days.

Testicular effects are described in section 7.5.2.

In the 90-day study, female mice had a dose-related adrenal lesion, which consisted of prominent cellular vacuolation in the zona reticularis contiguous with the medulla. This was apparent at the lowest concentration (Hamm, 1984).

No clear NOEL was established in the 2-week inhalation study, as (in addition to the testicular effects described in section 7.5.2 and the haematological effects described in section 7.7) increased relative kidney weights were observed in male Fischer rats at the lowest dose studied (51 mg/m^3) (Medinsky & Irons, 1985).

Similarly, no clear NOEL was established in the 90-day inhalation study, as female mice at the lowest dose showed cellular vacuolation of the adrenal gland (and haematological effects were observed in Fischer rats, as described in section 7.7) (Hamm, 1984).

7.2.4   Subcutaneous

Subcutaneous daily injection of adult male rabbits with 0.05 ml nitrobenzene/kg of body weight resulted in histopathological changes in the optic nerve, seen at necropsy at the end of the second week; changes included Marchi granules and regressive glial cell changes, with demyelinization seen as the 12-week study progressed (Yoshida, 1962).

7.3   Long-term toxicity and carcinogenicity

No studies were located regarding long-term toxicity or carcinogenic effects in animals after oral or dermal exposure to nitrobenzene.

The chronic toxicity and potential carcinogenicity of inhaled nitrobenzene have been evaluated following a 2-year exposure period in B6C3F$_1$ mice, Fischer-344 rats and Sprague-Dawley rats (CIIT, 1993). This data set was later analysed, summarized and discussed by CIIT scientists (Cattley et al., 1994). Subsequently, an extended analysis in the light of nitrobenzene's metabolism and potential mode of carcinogenic mode of action was presented (Holder, 1999a).

Male and female B6C3F$_1$ mice (70 per sex per dose group)[1] were exposed to 0, 26, 130 or 260 nitrobenzene/m^3 (0, 5, 25 or 50 ppm), whereas Fischer-344 rats (70 per sex per dose) and male Sprague-Dawley (CD) rats (70 per dose) were exposed to lower doses of 0, 5, 26 or 130 mg nitrobenzene/m^3 (0, 1, 5 or 25 ppm) because of the sensitivity of the rat to methaemoglobin formation. All exposures were for 6 h per day, 5 days per week, for a total of 505 exposure-days over a duration of 2 years.

Haematological findings from these studies are described in section 7.7, and effects on reproductive organs, in section 7.5.1.

Survival was not adversely affected by chronic nitrobenzene inhalation exposure (Cattley et al., 1994). There were only mild exposure-related decreases in body weights.

For mice, prominent dose-related changes included bronchiolization of alveolar walls (alveolar epithelium changed from a simple squamous to a tall columnar epithelium resembling that of the terminal bronchioles); centrilobular hepatocytomegaly and multinucleated hepatocytes; pigment deposition in and degeneration of olfactory epithelium; follicular cell hyperplasia in the thyroid; and increased nasal secretion. Compound-related findings seen at 260 mg/m^3 only included glandularization of the respiratory epithelium in the nose and bone marrow hypercellularity. Other compound-related findings (investigated in high-dose animals only) included thymic involution (females), mononuclear cell infiltration of the pancreas (females), kidney cysts (males) and hypospermia in the epididymis and diffuse testicular atrophy. For rats, prominent dose-related changes included eosinophilic foci in the liver (Fischer-344 rats), centrilobular hepatocytomegaly (Fischer-344 and CD males) and pigment deposition in olfactory epithelium. Spongiosis hepatitis was largely confined to the high-dose animals (both strains), as was kidney tubular hyperplasia (Fischer-344 males).

---

[1] Bioassay details have been reported by the CIIT (1993). The original protocol called for 60 test animals per sex per dose group, and 10 additional animals per group were to be set aside for interim sacrifice; however, CIIT investigators changed the protocol at 1 year and added the 10 to the 60, providing 70 animals per group (Cattley et al., 1994).

Tumour incidences from nitrobenzene inhalation were observed at eight different organ sites among the mice and two strains of rats (Tables 10–12). Each organ site presented indicates a significantly (i.e., biologically and statistically) positive carcinogenic response in the test animals arising from 2 years of exposure. Because there were no significant numbers of early deaths in any of the groups, the tumours appearing in the tables are those of aging animals.

B6C3F$_1$ male mice responded with alveolar and bronchial lung tumours (Table 10). The benign alveolar and bronchial tumours show a definite increase in trend. The malignant tumours, on the other hand, do not show a trend *per se*, but they do indicate that treated mice (i.e., 26, 130 and 260 mg/m^3) have more malignant alveolar and bronchial cancers on average (13%) than do the concurrent control mice (6%). Thyroid follicular cell adenomas (benign) were also increased in male B6C3F$_1$ mice (Table 10), whereas females showed no such trend. For B6C3F$_1$ female mice, malignant mammary gland tumours were increased in the 260 mg/m^3 group, but the 26 and 130 mg/m^3 groups were not analysed histologically for mammary tumours (Table 10) (CIIT, 1993; Cattley et al., 1994).

For male Fischer-344 rats, there were increases in both benign and malignant liver tumours (Table 11). Male Fischer-344 rats also responded with increased follicular cell benign and malignant thyroid cancers (Table 11). The third response site in Fischer-344 male rats was ostensibly a benign kidney response (Table 11). It is notable, however, that there was a single kidney adenocarcinoma in the high-dose group and none in the control group. In female Fischer-344 rats, benign uterine tumours (endometrial polyps) were observed. Because the concurrent control group has as many as 16% polyps, the effect may be an exacerbation of an age-related effect.

In Sprague-Dawley rats, liver cancers were seen (Table 12). The hepatic cancers were significant only because of the response at the high-dose level of 130 mg/m^3.

On the basis of non-cancer end-points, no NOELs were established in any of these studies (CIIT, 1993; Cattley et al., 1994). Findings at the lowest dose of 26 mg/m^3 in mice included lower body weights in females (weeks 16–30); several clinical chemistry changes,

Table 10. Incidence of tumours in B6C3F$_1$ mice[a]

	0 ppm	5 ppm	25 ppm	50 ppm
Male lung (alveolar/bronchial) adenomas or carcinomas				
Adenomas, alveolar/bronchial	7/68 (10%)	12/67 (18%)	15/65 (23%)	18/66 (27%)
Carcinomas, alveolar/bronchial	4/68 (6%)	10/67 (15%)	8/65 (12%)	8/66 (12%)
Total cancer incidence, alveolar/bronchial	9/68 (13%)	21/67 (31%)	21/65 (32%)	23/66 (35%)
Statistical results	trend = 0.017[b]	0.01[c]	0.007	0.003
Male thyroid follicular cell adenomas				
Adenomas, thyroid	0/65 (0%)	4/65 (6%)	1/65 (2%)	7/64 (11%)
Statistical results	trend = 0.015[b]	0.06[c]	0.5	0.006
Female mammary gland adenocarcinomas				
Mammary incidence	0/48 (0%)	not examined	not examined	5/60 (8%)
Statistical results	N/A[d]	–	–	0.049

[a] From Holder (1999a). B6C3F$_1$ mouse tumour incidences are the tumour occurrences divided by animals reported at risk plus animals carried over from interim sacrifice into the main CIIT study (Cattley et al., 1994). All mouse tumours occurred late in the 2-year inhalation bioassay study, and none of the interim-kill rodents indicated carcinogenicity at these sites (CIIT, 1993; Cattley et al., 1994). 1 ppm = 5.12 mg/m^3.

[b] The Peto trend probability P is given for each line of dose–response under the control column.

[c] The statistical probability of each pairwise response difference compared with the control incidence is presented under each dose column. The P value is estimated by Fisher's Exact Test.

[d] Not applicable.

Table 11. Incidence of tumours in Fischer-344 rats[a]

	0 ppm	1 ppm	5 ppm	25 ppm
Male hepatocellular adenomas/carcinomas				
Adenomas	1/69 (1%)	3/69 (4%)	3/70 (4%)	15/70 (21%)
Carcinomas	0/69 (0%)	1/69 (1%)	2/70 (3%)	4/70 (6%)
Total cancer incidence	1/69 (0%)	4/69 (6%)	5/70 (7%)	16/70 (23%)
Statistical results	trend = 0.001[b]	0.183[c]	0.108	<0.001
Male thyroid follicular cell adenomas/adenocarcinomas				
Adenomas	0/69 (0%)	0/69 (0%)	2/70 (3%)	2/70 (3%)
Adenocarcinomas	2/69 (3%)	1/69 (1%)	3/70 (4%)	6/70 (9%)
Total cancer incidence	2/69 (3%)	1/69 (1%)	5/70 (7%)	8/70 (11%)
Statistical results	trend = 0.010[b]	0.500 N[c,d]	0.226	0.051
Male kidney tubular adenomas/adenocarcinomas				
Adenomas	0/69 (0%)	0/68 (0%)	0/70 (0%)	5/70 (7%)
Carcinomas	0/69 (0%)	0/68 (0%)	0/70 (0%)	1/70 (1%)
Total incidence	0/69 (0%)	0/68 (0%)	0/70 (0%)	6/70 (9%)
Statistical results	trend < 0.001[b]	–[e]	–[e]	0.015
Female endometrial polyps				
Overall uterine incidence	11/69 (16%)	17/65 (26%)	15/65 (23%)	25/69 (36%)
Statistical results	trend = 0.01[b]	0.107[c]	0.205	0.006

Table 11 (Contd).

^a From Holder (1999a). Fischer-344 rat tumour incidences are the tumour occurrences divided by animals reported at risk plus animals carried over from interim sacrifice into the main CIIT study (Cattley et al., 1994). All rat tumours occurred late in the 2-year inhalation bioassay study, and none of the interim-kill rodents indicated carcinogenicity at these sites (CIIT, 1993; Cattley et al., 1994). 1 ppm = 5.12 mg/m³.

^b The Peto trend probability P is given for each line of dose–response under the control column.

^c The statistical probability of each pairwise response difference compared with the control incidence is presented under each dose column. The P value is estimated by Fisher's Exact Test.

^d N = no statistical (or null) effect.

^e A "–" means that no sampling of tissues was made for histopathology by CIIT pathologists.

Table 12. Incidence of tumours in Sprague-Dawley (CD) rats[a]

	0 ppm	1 ppm	5 ppm	25 ppm
Male hepatocellular adenomas or carcinomas				
Overall liver incidence	2/63 (3%)	1/67 (1%)	4/70 (6%)	9/65 (14%)
Statistical results	trend = 0.002[b]	0.447 N[c,d]	0.391	0.031

[a] From Holder (1999a). Sprague-Dawley rat tumour incidences are the tumour occurrences divided by animals reported at risk plus animals carried over from interim sacrifice into the main CIIT study (Cattley et al., 1994). All rat tumours occurred late in the 2-year inhalation bioassay study, and none of the interim-kill rodents indicated carcinogenicity at these sites (CIIT, 1993; Cattley et al., 1994). 1 ppm = 5.12 mg/m³.

[b] The Peto trend probability P is given for each line of dose–response under the control column.

[c] The statistical probability of each pairwise response difference compared with the control incidence is presented under each dose column. The P value is estimated by Fisher's Exact Test.

[d] N = no statistical (or null) effect.

102

either statistically significant at this dose or showing a dose-related trend; and reduced weights (absolute and relative) of liver and kidney. Histopathological findings at the low dose included an increased incidence of secretory product in respiratory epithelial cells; degeneration/loss of olfactory epithelium; dilation of submucosal glands and accumulation of brown pigment-containing macrophages in the submucosal areas of olfactory epithelium; bronchiolization of alveolar cell walls; hepatocytomegaly and multinucleated hepatocytes; and bone marrow hypercellularity. In rats, findings at the lowest dose of 5 mg/m^3 included slight but significant decreases in serum sodium. Histopathological findings at the low dose included accumulation of brown pigment-containing macrophages in the mucosa and submucosa of olfactory epithelium; inflammation in the anterior nasal passages, including suppurative exudate; and mucosal epithelial hyperplasia (CD rats). For haematological toxicity and effects on fertility, see sections 7.7 and 7.5.1, respectively.

7.4 Mutagenicity and related end-points

The genotoxicity of nitrobenzene has been tested in an array of non-mammalian and mammalian test systems *in vitro* and *in vivo*. The results have usually been negative. In a few studies, methods were used in which the genetic basis of the effects observed is not fully understood; thus, the relevance of these studies is not clear. Information on the genetic activity of nitrobenzene is comprehensively reviewed below.

7.4.1 DNA interactions

In five separate experiments using hepatocytes prepared from different human livers, nitrobenzene at concentrations up to 1 mmol/litre was negative for induction of unscheduled DNA repair (Butterworth et al., 1989). In parallel assays using rat liver primary hepatocyte cultures, nitrobenzene was similarly negative.

Radioactivity, most probably due to covalent binding to DNA, was observed in DNA isolated from rat liver and kidney and mouse liver and lung samples collected 24 h after a single subcutaneous injection of ^{14}C-labelled nitrobenzene at 4 mg/kg of body weight (BASF, 1997). Radioactivity, expressed as covalent binding indices, was at the upper end of the range of values typically found with weak genotoxic carcinogens.

In an *in vivo–in vitro* hepatocyte DNA repair test, rats were gavaged with nitrobenzene at 200 or 500 mg/kg of body weight and killed after 12 h; their livers were then removed, and primary cultures of hepatocytes were incubated with [³H]thymidine. There was no evidence of any increase in unscheduled DNA synthesis (Mirsalis et al., 1982).

7.4.2 Mutation

Results of *in vitro* gene mutation studies in bacteria are summarized in Table 13. As indicated in the table, a number of independent studies did not find any significant mutagenic activity with nitrobenzene in the *Salmonella typhimurium* histidine reversion test, either in the presence or in the absence of S9 mix, at concentrations up to 3300 μg/plate (Garner & Nutman, 1977; Anderson & Styles, 1978; Chiu et al., 1978; Purchase et al., 1978; Ho et al., 1981; Haworth et al., 1983; Suzuki et al., 1983; Hughes et al., 1984; Nohmi et al., 1984; Vance & Levin, 1984; Kawai et al., 1987; Dellarco & Prival, 1989).

The non-mutagenicity of nitrobenzene in bacterial gene mutation tests was further confirmed in studies sponsored under contract by the US NTP (unpublished work cited in Beauchamp et al., 1982). Nitrobenzene was also found to be without mutagenic potential in *Salmonella typhimurium* (TA100-FR50) deficient in oxygen-insensitive nitroreductase activity (Benkendorf, 1978); the bacteria were treated with nitrobenzene under anaerobic conditions to allow reduction by the oxygen-sensitive enzyme. Suzuki et al. (1983) confirmed the non-mutagenicity of nitrobenzene in nitroreduction-competent tester strains of *Salmonella* even with S9 addition, although co-mutagenicity of nitrobenzene with the co-mutagen norharman (found in the pyrolysate of tryptophan and in tobacco smoke) was observed. Similarly, in a modified Ames test utilizing a preincubation step and the addition of flavin mononucleotide to the hamster liver S9 mix to provide more general nitroreducing conditions, nitrobenzene at concentrations up to 10 μmol/plate (1200 μg/plate) tested negative in tester strains TA98 and TA100, even though significant nitroreduction occurred (Dellarco & Prival, 1989).

Kuroda (1986) examined the mutagenic effects of nitrobenzene in cultured Chinese hamster lung (V79) cells; the induction of 8-azaguanine, 6-thioguanine and ouabain resistance was determined. Nitrobenzene, without metabolic activation, induced 8-azaguanine resistance with low frequency at 0.6 μg/ml, an effect enhanced when S9 was added to the medium. The frequency of ouabain resistance was marginally increased, but only in the absence of S9.

Table 13. Summary of results from *Salmonella typhimurium* (histidine reversion) tests[a]

Strain	Concentration[b] (µg/plate)	–S9	+S9	Reference
TA1538	100	–	–	Garner & Nutman, 1977
TA98, TA100, TA1535, TA1538	2500	ND	–	Anderson & Styles, 1978
TA98, TA100, TA1535, TA1537	NS	ND	–	Purchase et al., 1978
TA98, TA100	1230[c]	–	ND	Chiu et al., 1978
TA100-FR50[d]	NS	ND	–	Benkendorf, 1978
TA98	500	ND	–	Ho et al., 1981
TA98	100	–	+[e]	Suzuki et al., 1983
TA100	100	–	ND	Suzuki et al., 1983
TA98, TA100, TA1535, TA1537	NS	–	–[f]	Haworth et al., 1983
TA97a,[g] TA98, TA98NR,[h] TA100, TA100NR, TA1535, TA1537, TA1537NR, TA1538	1000	–	ND	Vance & Levin, 1984
TA98, TA100	NS	–	–	Nohmi et al., 1984
TA97, TA98, TA100	3300	–	–	Hughes et al., 1984
TA98, TA100	NS	–	–	Kawai et al., 1987
TA98, TA100	1230[c]	–	–	Dellarco & Prival, 1989
TA98, TA100	316–2525	–	–	Aßmann et al., 1997

[a] – = negative result; + = positive result; ND = no data; NS = concentration not stated; –S9 = without activation; +S9 = with activation; in most cases, S9 fractions were prepared from livers of rats (induced with Aroclor-1254, Kanechlor KC-400 or phenobarbitone).
[b] Only the highest (non-toxic) concentration tested is listed here.
[c] Stated concentration was 10 µmol/plate.
[d] Limited nitroreductase activity even in the presence of air.
[e] Only in the presence of the co-mutagen norharman; negative in its absence.
[f] Haworth et al. (1983) utilized S9 from both rat and hamster livers.
[g] TA97 isolate with a defined uvrB deletion.
[h] NR = nitroreductase-deficient strain.

There was apparently a weak increase of gender-linked recessive lethal mutations in *Drosophila* when nitrobenzene was added to the food, but a significant increase when *Drosophila* were exposed to nitrobenzene fumes over 8–10 days (Rapoport, 1965). In view of the paucity of details, this work provides little meaningful information.

7.4.3 Chromosomal effects

Huang et al. (1996) reported that nitrobenzene at 50 mmol/litre caused chromosomal aberrations in cultured human lymphocytes *in vitro*. Unfortunately, the results were reported only as positive or negative and at only one concentration — i.e., the lowest dose, which was claimed to be positive. The result was considered positive if there was a statistically significant increase ($P < 0.01$) above the negative control; however, the statistical test used was not stated. Additionally, no data were provided on the cytotoxicity, and the dose of nitrobenzene at which positive genotoxicity was reported was 5 times higher than the recommended upper test concentration for relatively non-cytotoxic compounds (OECD Test Guideline 473, adopted 21 July 1997).

Male Fischer-344 rats were exposed to concentrations of 0, 26, 82 and 260 mg/m³ (0, 5, 16 and 50 ppm) for 6 h per day, 5 days per week, over a 29-day period (21 exposures) via inhalation. Cultures of blood and isolated spleen lymphocytes were grown in the presence of 2 μmol 5-bromodeoxyuridine/litre. No effects on sister chromatid exchange frequency or on chromosomal aberrations (excluding gaps) were observed. However, nitrobenzene did have a significant inhibitory effect on the mitotic activity and cell cycle progression of concanavalin A-stimulated peripheral blood lymphocytes (Kligerman et al., 1983).

Male and female B6C3F₁ mice were given a single intraperitoneal injection of 62.5, 125 or 250 mg/kg of body weight of nitrobenzene. No increase in micronucleated polychromatic erythrocytes was observed at 24 or 48 h after the treatment in the bone marrow of the animals (BASF et al., 1995).

7.4.4 Cell transformation

In an assay examining mammalian cell transformation in culture, baby Syrian hamster kidney cells (BHK-21 C13) and human diploid lung fibroblasts (WI-38) were exposed to nitrobenzene (final test concentrations ranged from 0.08 µg/ml to 250 mg/ml) in the presence and absence of rat liver S9 fraction. Nitrobenzene at the LC_{50} (which was not stated) did not cause cell transformation in the cell systems used (Styles, 1978).

7.4.5   *Genotoxicity of nitrobenzene metabolites*

The genotoxicity of eight putative nitrobenzene metabolites — i.e., nitrosobenzene, *N*-phenylhydroxylamine (hydroxylaminobenzene), *p*-nitrosophenol, *p*-nitrophenol, *p*-aminophenol, acetanilide, *p*-hydroxyacetanilide and aniline — was summarized in a review by Beauchamp et al. (1982).

The urinary metabolites of nitrobenzene in rats and mice — *p*-hydroxyacetanilide, *p*-nitrophenol, *m*-nitrophenol and *p*-aminophenol, and their sulfate conjugates — have given negative results in genotoxicity tests (McCann et al., 1975; Bartsch et al., 1980; Wirth et al., 1980; Probst et al., 1981; Wilmer et al., 1981; see also Beauchamp et al., 1982).

There is some evidence of the genetic activity of some of the putative metabolites of nitrobenzene, although most of the data are negative. Ohkuma & Kawanishi (1999) investigated the mechanism of DNA damage induced by nitrosobenzene (in calf thymus DNA *in vitro*). The authors reported that nitrosobenzene can be reduced non-enzymatically by NADH, and the redox cycle reaction resulted in oxidative DNA damage due to the copper–oxygen complex, derived from the reaction of copper(I) with hydrogen peroxide. Aniline induced gene mutations in Chinese hamster lung (V79) cells (Kuroda, 1986) and in mouse L5178Y cells (Amacher et al., 1980; Caspary et al., 1988; Wangenheim & Bolcsfoldi, 1988), sister chromatid exchanges (Abe & Sasaki, 1977; Cunningham & Ringrose, 1983; Galloway et al., 1987) and chromosomal aberrations (Galloway et al., 1987) in Chinese hamster cells *in vitro* and morphological transformation of BALB/c 3T3 cells (Dunkel et al., 1981). It also induced DNA damage in the liver and kidney of rats and sister chromatid exchanges in the bone marrow of mice (Parodi et al., 1981, 1982). Aniline, a reductive metabolite of nitrobenzene (tested at up to 1 mmol/litre in six separate assays), was negative for unscheduled DNA repair in primary cultures of rat and human hepatocytes (Butterworth et al., 1989). The draft European Union (EU) risk assessment report on aniline (EU, 2001) provides further evidence on the genotoxicity of aniline. *p*-Aminophenol induced gene mutations in *Salmonella typhimurium* strain TA1535 and was also reported to be positive in the mouse *in vivo* micronucleus test (Wild et al., 1980). *p*-Nitrophenol caused DNA damage in *Bacillus subtilis* (Shimizu & Yano, 1986) and *Proteus mirabilis* (Adler et al., 1976). *p*-Nitrosophenol was weakly mutagenic to *Salmonella typhimurium* strain TA1538 (Gilbert et al., 1980).

Aßmann et al. (1997) reported that substances derived from nitrobenzene or aniline by addition of at least one nitro group in the *meta* or *para* position were mutagenic in the Ames test (strains TA98 and TA100). Nitrobenzene and aniline themselves were non-mutagenic.

7.5 Reproductive toxicity

7.5.1 Effects on fertility

7.5.1.1 Oral

Following a single oral dose of 300 mg/kg of body weight, findings included degeneration of the seminiferous epithelium of the testes within 3 days of treatment and an approximately 17-day period of aspermia after a 17- to 20-day lag period. Histological examination showed that pachytene spermatocytes and step 1–2 spermatids were the most susceptible cell stages. Repair was substantial by 3 weeks after treatment, with >90% regeneration of seminiferous epithelium by 100 days after treatment (Levin et al., 1988).

Nine-week-old Sprague-Dawley male rats were gavaged with 60 mg nitrobenzene/kg of body weight (in sesame oil) for up to 70 days; on days 7, 14, 21, 28, 42, 56 and 70 of treatment, they were mated with normal proestrous females (Kawashima et al., 1995b). On the day after each of these matings, the five males and five non-mating males in each group (control and treated) were sacrificed for morphological examination and sperm assessment. After 7 days of treatment, there were no effects on sperm motility, progressive sperm motility (movement along a capillary tube), sperm count, sperm morphology and viability, or testicular and epididymal weights. By day 14, sperm motility, progressive sperm motility, sperm count and testicular and epididymal weights were significantly decreased. By day 21, sperm viability and the fertility index also decreased (the latter to about 15%

of control), with increases in abnormal sperm and instances in which no motile sperm were evident. By day 28, the fertility rate was 0%. However, even after 70 days, the copulation index was unaffected at this dose. The most sensitive spermatic end-points were determined to be sperm count and sperm motility, followed by progressive motility, viability, presence of abnormal sperm and, finally, the fertility index.

In the Mitsumori et al. (1994) ReproTox study, described in section 7.2.1, significant decreases were observed in testes and epididymis weights in mid- and high-dose groups. Histopathologically, treated males showed atrophy of the seminiferous tubules of the testes, with dose-dependent incidence (0/10, 1/10, 10/10 and 10/10 in the 0, 20, 60 and 100 mg/kg of body weight per day dose groups, respectively) and severity (not stated). In addition, there was Leydig cell hyperplasia (respective incidence of 0/10, 0/10, 10/10 and 8/10), decreased numbers of cells with round nuclei per seminiferous tubule (incidence not stated) and loss of intraluminal sperm in the epididymis (0/10, 0/10, 10/10 and 10/10, respectively).

Male fertility was not affected. The body weights of pups from treated dams were lowered, and postnatal loss was increased. Otherwise, there were no obvious effects on copulation, fertility, implantation indices or gestation period length. The lack of effect of nitrobenzene on male fertility may be due to the short premating dosing interval used in the study and to the fact that rats produce sperm in very large abundance.

In the 13-week NTP study (see section 7.2.1), testicular atrophy was observed in mice at 18.75 mg/kg of body weight per day (3/10 animals), 37.5 mg/kg of body weight per day (2/10), 150 mg/kg of body weight per day (5/10) and 300 mg/kg of body weight per day (5/10). In Fischer-344 rats, the testes were mildly to markedly atrophic at the two highest doses, with varying degrees of hypospermatogenesis and multinucleated giant cell formation.

Morrissey et al. (1988) reviewed the reproductive organ toxicity of nitrobenzene in B6C3F$_1$ mice and Fischer-344 rats, based on the results of a number of 13-week studies conducted by the US NTP. In gavage studies with nitrobenzene in Fischer-344 rats (9.4, 37.5 and 75 mg/kg of body weight) and B6C3F$_1$ mice (18.75, 75 and 300 mg/kg of body weight), no significant effects were noted on body weights;

however, the weights of the right cauda, epididymis and testes and the motility and density of sperm were significantly decreased in one or more treatment groups. In addition, there was an increase in the percentage of abnormal sperm.

7.5.1.2   *Dermal*

In the range-finding NTP study (see section 7.2.2), nitrobenzene was administered to $B6C3F_1$ mice and Fischer-344 rats at doses of 200–3200 mg/kg of body weight per day for 14 days (NTP, 1983b). Histologically, mice and rats showed changes in testes, with mice less affected than rats.

In the 13-week NTP study (see section 7.2.2), testicular atrophy was seen in all mice at the high dose of 800 mg/kg of body weight per day. Uterine atrophy was seen in one, one and five mice in the 200, 400 and 800 mg/kg of body weight per day dose groups, respectively. In Fischer-344 rats, the testes were moderately to markedly atrophic at the two highest doses (400 and 800 mg/kg of body weight per day), with varying degrees of hypospermatogenesis and multinucleated giant cell formation.

Morrissey et al. (1988) reviewed the reproductive organ toxicity of nitrobenzene in $B6C3F_1$ mice and Fischer-344 rats after dermal exposure, based on the results of a number of 13-week studies conducted by the US NTP. In these dermal studies in Fischer-344 rats at 50, 200 and 400 mg/kg of body weight, weights of the right cauda, epididymis and testis and the motility and density of sperm were significantly decreased; at odds with the expected results, there was an apparent decrease in abnormal sperm, but the testicular effects were so severe that few sperm were available for analysis. In $B6C3F_1$ mice at the same dermal doses, there was an observed increase in abnormal sperm, but otherwise the only change was a decrease in sperm motility.

7.5.1.3   *Inhalation*

Dodd et al. (1987) exposed 38- to 42-day-old Sprague-Dawley rats (30 per sex per group) to nitrobenzene via the inhalational route, at concentrations of 0, 5, 51 or 200 mg/m³ (0, 1, 10 or 40 ppm), in a two-generation reproduction study (1:1 matings). Exposures were for

6 h per day, 5 days per week, for 10 weeks prior to mating, during mating and through day 19 of gestation. On postnatal day 5, dams were removed from their litters for the exposure period, remaining with their pups for the rest of the time. Necropsy occurred on or just after day 21. F_1 rats (30 per sex per group selected) were allowed a 2-week growth period prior to nitrobenzene exposure, with the subsequent exposure regime the same as for the F_0 generation; some F_1 males were not sacrificed after mating, but were allowed a recovery period. F_2 pups were sacrificed on postnatal day 21. No nitrobenzene-related effects on reproduction were noted at either 5 or 51 mg/m³. At 200 mg/m³, there was a decrease in the fertility index (number of pregnancies/number of females mated) of F_0 and F_1 generations, associated with alterations in male reproductive organs — i.e., reduced weights of testes and epididymis, seminiferous tubule atrophy, spermatocyte degeneration and the presence of giant syncytial spermatocytes. Fertility indices were 30/30, 27/30, 29/30 and 16/30 in the respective F_0 groups and 30/30, 27/30, 26/30 and 3/30 in the respective F_1 groups. The only significant finding in litters from rats exposed to 200 mg/m³ was a decrease in the mean body weight of F_1 pups on postnatal day 21. Survival indices were unaltered. After a 9-week recovery period, F_1 males from the 200 mg/m³ group were mated with 77-day-old untreated females in order to examine the reversibility of nitrobenzene effects on the gonads. An increase in the fertility index (above that measured during nitrobenzene exposure) indicated at least partial functional reversibility upon removal from nitrobenzene exposure; fertility indices in the control and high-dose recovery groups were 29/30 and 14/30, respectively. The numbers of giant syncytial spermatocytes and degenerated spermatocytes were greatly reduced, but testicular seminiferous tubule atrophy persisted, although active stages of spermatocyte degeneration were much less frequent. Maternal toxicity was not observed. Under the conditions of this study (i.e., exposure for 6 h per day, 5 days per week), a NOEL of 51 mg/m³ was established for reproductive toxicity in Sprague-Dawley rats.

In the 2-week CIIT study (Medinsky & Irons, 1985), described in section 7.2.3, a very prominent decrease in relative testicular weights was evident at the highest dose in Fischer-344 rats, a finding that showed no recovery by day 14. Testicular lesions in Fischer-344 rats exposed for 2 weeks at 640 mg/m³ consisted of increased multi-nucleated giant cells, Sertoli cell hyperplasia and severe dysspermio-genesis, with maturation arrested at the level of primary and secondary spermatocytes; the epididymis contained reduced numbers of mature sperm, a finding still evident after 2 weeks of recovery. Dysspermiogenesis of moderate severity was seen in CD rats exposed at 640 mg/m³. The testes of mice exposed at 640 mg/m³ showed a different lesion, with acute testicular degeneration, an absence of spermatozoa in seminiferous tubules and the epididymis and degeneration of tubular epithelial cells (Medinsky & Irons, 1985).

In the 13-week CIIT study (Hamm, 1984; Hamm et al., 1984) described in section 7.2.3, both rat strains had reduced testicular weights, bilateral degeneration of seminiferous epithelium and a reduction or absence of sperm in the epididymis at 260 mg/m³; CD rats appeared more severely affected, with all 260 mg/m³ animals showing gross bilateral testicular atrophy. Marginal effects were noted in CD rats exposed at 82 mg/m³. Mice had no testicular lesions at these doses (Hamm, 1984; Hamm et al., 1984).

No NOELs for reproductive effects were established in the 2-year CIIT studies. Findings at the lowest dose of 26 mg/m³ in mice included diffuse testicular atrophy; in rats, findings at the lowest dose of 5 mg/m³ included benign uterine endometrial stromal polyps (Fischer-344) (see section 7.3). Histopathological findings at the low dose included abnormal sperm (Fischer-344 rats) and increased bilateral atrophy (CD rats).

7.5.2   Testicular toxicity

7.5.2.1   In vitro *studies*

Allenby et al. (1991) studied the effect of nitrobenzene on the *in vitro* secretion of immunoactive inhibin (basal and stimulated) by cultured isolated rat seminiferous tubules and by co-cultures of immature rat Sertoli cells or Sertoli cells plus germ cells, as well as on *in vivo* secretion in Sprague-Dawley rats. Nitrobenzene at 10 μmol/litre or 1 mmol/litre stimulated basal secretion of inhibin in seminiferous tubule cultures but did not affect secretion stimulated by follicle stimulating hormone (FSH) or dibutyryl cyclic AMP. It also enhanced secretion by Sertoli cell cultures, but to a lesser extent. Exposure *in vivo* to a single dose of nitrobenzene (300 mg/kg of body weight by gavage in corn oil) resulted in 2- to 4-fold increases in inhibin levels in testicular interstitial fluid at 1 and 3 days post-treatment, associated with early impairment of spermatogenesis, as judged by testicular weight.

The effects of nitrobenzene on Sertoli cells were assessed *in vitro* using Sertoli cell and Sertoli cell plus germ cell co-cultures (Allenby et al., 1990). Gross morphological changes, including vacuolation of Sertoli cells, were observed following treatment of cultures with 1 mmol nitrobenzene/litre. Exposure of co-cultures to nitrobenzene also resulted in dose-dependent exfoliation of predominantly viable germ cells. Nitrobenzene (>500 µmol/litre) significantly stimulated the secretion of lactate and pyruvate by Sertoli cells, an effect that was more marked in the absence of germ cells. Comparable changes were observed in FSH-stimulated cultures. Inhibin secretion by Sertoli cells was also altered by exposure to nitrobenzene, but in a biphasic manner: low (10 nmol/litre to 1 µmol/litre) and high (100 µmol/litre to 1 mmol/litre) doses enhanced inhibin secretion, whereas intermediate (10 µmol/litre) doses had no effect. These effects were evident in both culture systems, but inhibin secretion by Sertoli cell plus germ cell co-cultures was always greater than that by Sertoli cell cultures. The effects of nitrobenzene on inhibin secretion were not evident in FSH-stimulated cultures.

The effects of nitrobenzene on protein secretion by seminiferous tubules isolated from rats were assessed. Seminiferous tubules were isolated from immature (28-day), late pubertal (45-day) and young adult (70-day) rats and cultured *in vitro* for 24 h with [^{35}S]methionine in the presence or absence of 100 µmol nitrobenzene/litre. Incorporation of [^{35}S]methionine into newly synthesized proteins in the culture medium (secreted proteins) was assessed using two-dimensional sodium dodecylsulfate polyacrylamide gel electrophoresis. Nitrobenzene *in vitro* had no effect on the incorporation of [^{35}S]methionine into overall secreted proteins by seminiferous tubules isolated from immature rats, whereas addition of nitrobenzene to immature rat Sertoli cell plus germ cell co-cultures resulted in increased incorporation of radiolabel into secreted proteins. In contrast, the same additions to seminiferous tubules isolated from adult rats resulted in a 20–34% decrease in the overall incorporation of [^{35}S]methionine. Seminiferous tubules isolated from late pubertal rats showed a response similar to that of seminiferous tubules from adult rats, except that the decreases in incorporation were smaller. Electrophoretic analysis revealed considerable age-dependent differences in the proteins secreted by seminiferous tubules from immature and adult rats; most of these proteins were prominent secretory products of seminiferous tubules from adult rats, but were minor or non-detectable products of cultures of seminiferous tubules or Sertoli cells plus germ cells from immature rats. Most disappeared or decreased in abundance after culture of seminiferous tubules with nitrobenzene (McLaren et al., 1993a). It appears that germ cells modulate the secretory function of Sertoli cells and that protein secretion by Sertoli cells in immature and adult animals is differentially affected by nitrobenzene exposure. Further work indicated that exposure to nitrobenzene caused stage-specific changes in the secretion of proteins by isolated seminiferous tubules (McLaren et al., 1993b); studies in rats have identified proteins in blood that derive from Sertoli cells and germ cells.

In *ex vivo* studies on the effects of nitrobenzene on the secretion of proteins by isolated seminiferous tubules, adult rats received single oral doses of 300 mg nitrobenzene/kg of body weight. Long lengths of seminiferous tubules at different stages of the spermatogenic cycle (i.e., II–V, VI–VIII or IX–XII) were then isolated from control and treated rats at 1 or 3 days post-treatment and cultured *in vitro* for 24 h with [^{35}S]methionine. Incorporation of [^{35}S]methionine into secreted proteins was assessed and the pattern of protein secretion evaluated using two-dimensional sodium dodecylsulfate polyacrylamide gel electrophoresis. Seminiferous tubules isolated from rats pretreated 24 h earlier with nitrobenzene *in vivo* showed a significant decrease in the overall incorporation of [^{35}S]methionine into secreted proteins at stages VI–VIII and IX–XII, not at stages II–V. In similar *in vitro* experiments, seminiferous tubules at the same stages were isolated from untreated rats and cultured in the presence or absence of 100 μmol nitrobenzene/litre for 24 h. Comparable protein changes were observed as in the *ex vivo* experiments (McLaren et al., 1993b).

7.5.2.2   In vivo *studies*

A study that evaluated sperm viability using two fluorescent pigments, Calcein AM (which permeates intact cell membranes and indicates intracellular esterase activity) and ethidium homodimer (which permeates impaired cell membranes and combines with nucleic acids), showed that nitrobenzene given to rats at 20, 40 or 60 mg/kg of body weight (dose route not stated) produced marked effects on sperm numbers, motility and survival (Kato et al., 1995).

Nitrobenzene at 300 mg/kg of body weight elicited expected histopathological responses after a single gavage dose (in corn oil) in Sprague-Dawley rats. The main histopathological effects seen were degenerating spermatocytes (degenerating and missing pachytene spermatocytes in stages VII–XIV) at 2 days after treatment, with immature germ cells and debris in the initial segment of the epididymis. At day 14, maturation depletion of spermatids in stages V–XIV, some multinucleated giant cells and testicular debris throughout the epididymis were reported (Linder et al., 1992).

Analysis of sperm from 10-week-old male Sprague-Dawley rats treated with nitrobenzene (60 mg/kg of body weight by the oral route) for up to 14 days using an image processor and motion analysis software found, in addition to reduced sperm density, significant decreases in straight-line distance and straight-line velocity, but not in curvilinear distance, curvilinear velocity or amplitude of lateral head displacement (Kawashima et al., 1995a).

Using a flow cytometric analysis, morphological changes in sperm after a single oral dose of 100 or 300 mg nitrobenzene/kg of body weight to 8-week-old Sprague-Dawley IGS rats were investigated at 28 days after treatment. The study revealed an increase of 38.5% in abnormal sperm, with a 56.8% increase in the incidence of tailless sperm. In addition to tailless sperm, incidences of sperm with abnormalities such as no-hook, banana and pin shapes were slightly increased (Yamamoto et al., 2000).

The possible involvement of apoptosis in the process of rat germ cell degeneration caused by nitrobenzene was examined (Shinoda et al., 1998). Adult Sprague-Dawley rats were treated with a single oral dose of nitrobenzene (250 mg/kg of body weight) and killed at 6, 12, and 24 h and 2, 3, 5 and 7 days. The earliest morphological signs of germ cell degeneration in testes were found in pachytene spermatocytes 24 h after dosing. In degenerating spermatocytes, marked nuclear chromatin condensation at the nuclear periphery and crowding of cytoplasmic constituents, signs characteristic of apoptosis, were observed. Degenerating spermatocytes contained fragmented DNA. The presence of DNA laddering on electrophoresis gels, a hallmark of apoptosis, was first apparent and most prominent at 24 h, gradually becoming less detectable. No such changes were observed up to 12 h after dosing or in control animals. Thus, apoptotic mechanisms were demonstrated in the induction of spermatocyte degeneration caused by nitrobenzene.

Cytotoxic effects of nitrobenzene on spermatogenesis in the testes of mature Sprague-Dawley (Crj:CD) rats were analysed by measuring the DNA content distribution and testicular weights at 1, 2 and 3 weeks after daily oral doses of 60 mg nitrobenzene/kg of body weight (Iida et al., 1997). Within a week of administration, a large number of 1C cells (cells of the ploidy compartment 1C) were lost and meiosis of secondary spermatocytes was suppressed, but nitrobenzene had little effect on spermatocytes prior to the early pachytene stage. The proportion of 1C cells returned to nearly normal during a 2-week recovery period.

7.5.3 *Embryotoxicity and teratogenicity*

Tyl et al. (1987) exposed pregnant Sprague-Dawley rats from day 6 to 15 of gestation, 6 h per day, to nitrobenzene vapour concentrations of 0, 5, 51 and 200 mg/m^3 (0, 1, 10 and 40 ppm). Maternal weight gain was significantly depressed during the dosing period at 200 mg/m^3, with full recovery by day 21. At necropsy on day 21, absolute and relative spleen weights were increased in the dams at 51 and 200 mg/m^3, but there were no treatment-related effects on gravid uterine weight, liver weight or kidney weight, on pre- or postimplantation loss, including resorptions and dead fetuses, on the sex ratio of live fetuses, on fetal body weights per litter or on the incidence of fetal malformations or variations. Thus, there was no developmental toxicity associated with inhaled nitrobenzene at concentrations that produced some maternal toxicity (51 and 200 mg/m^3).

In a range-finding teratology study, nitrobenzene (99.9%) was administered by the inhalational route (whole-body exposure in 10-m^3 chambers) to mated New Zealand White rabbits (12 per group) for 6 h per day during days 7–19 of gestation (Bio/dynamics Inc., 1983). Targeted levels were 0, 10, 40 and 80 ppm, with mean actual exposures of 0, 51, 200 and 410 mg/m^3 (0, 10, 40 and 81 ppm). Surviving females were necropsied on day 20. There were no adverse maternal effects on mortality, body weights, clinical observations or gross postmortem observations. Kidney and liver weights were not significantly affected. There were no significant effects on numbers of corpora lutea, implants, resorptions or fetuses. On days 13 and 19 of gestation, methaemoglobin levels were significantly higher than controls in the high-dose group, while on day 20, methaemoglobin levels were significantly higher than controls at the middle and high doses.

Nitrobenzene (99.8% purity) was administered by the inhalational route (whole-body exposure in 10-m^3 chambers) to mated New Zealand White rabbits (4–5 months old, 22 per group; from Hazleton Dutchland Inc., Pennsylvania, USA) for 6 h per day during days 7–19 of gestation (Bio/dynamics Inc., 1984). Targeted levels were 0, 10, 40 and 100 ppm, with mean analytical concentrations of 0, 51, 210 and 530 mg/m^3 (0, 9.9, 41 and 104 ppm). Dams were necropsied on day 30. There were no significant adverse effects on maternal mortality, body weights, clinical signs or gross postmortem observations. At 51 mg/m^3, nitrobenzene was not maternally toxic, embryotoxic or teratogenic. At 210 mg/m^3, there was some maternal toxicity, as indicated by an increase in methaemoglobin (40% over controls) and liver weight (11.5% increase in relative weight compared with controls), but there was no evidence of embryotoxicity or teratogenicity. At the highest exposure level, nitrobenzene was maternally toxic — i.e., slight body weight loss during the dosing period, methaemoglobinaemia (60% over control) and increased liver weight (11.9% increase in relative weight compared with controls) — and there was limited evidence of some embryotoxicity (i.e., a possible increase in resorptions), albeit not statistically significant. There was no induction of terata. This study was also reported in abstract form (Schroeder et al., 1986).

7.6 Skin and eye irritation and sensitization

The method of Draize was used to test nitrobenzene for eye irritation in male albino rabbits (2.5–3 kg of body weight; strain not stated) (Sziza & Magos, 1959). A volume of 0.05 ml introduced under the lower eyelid resulted in minimal effects, with reported scores of 8 and 3 at 1 and 24 h, respectively, with no findings (0 score) at 48 and 96 h. Skin irritation was also tested in male albino rabbits, using a volume of 0.05 ml. A score of 1 (barely perceptible or very slight erythema) was recorded at 24 h, with 0 scores at 48, 72 and 96 h. In male guinea-pigs (600–900 g of body weight), nitrobenzene (3% solution in acetone) did not cause skin sensitization.

A study investigating *in vitro* alternatives to the Draize test for eye irritation (Spielmann et al., 1991) found that nitrobenzene could be classified as a non-irritant according to the HET-CAM test, a test performed on the chorioallantoic membrane of hen eggs.

7.7 Haematological toxicity

7.7.1 Oral

Male B6C3F$_1$ and Swiss Webster mice and Fischer-344 rats were given nitrobenzene by gavage at 0, 150, 200 or 300 mg/kg of body weight per day for 3 days and sacrificed 24 h after the last dose (Goldstein et al., 1983b). Specific investigations on the spleen in mice revealed splenic lesions consisting of slight congestion, erythroid hyperplasia and lymphoid hyperplasia. In contrast, marked splenic congestion and comparatively little lymphoid hyperplasia were seen in the rats. A dose-related increase in macromolecular covalent binding of [^{14}C]nitrobenzene in spleen was seen in all animals. Rat splenic and erythrocytic binding was 6–10 and 2–3 times greater, respectively, than that in mice, correlating with the increased severity of congestion. Erythrocyte covalent binding was greater than splenic binding in both mice and rats.

Goldstein et al. (1984a, 1984b) studied the influence of dietary pectin, a fermentable carbohydrate, on intestinal microfloral metabolism and toxicity of nitrobenzene. Dietary pectin is known to alter the intestinal microfloral metabolism of some xenobiotics. Male Fischer-344 rats were fed, by gavage, a purified diet containing 5% cellulose, a purified diet with 5% pectin replacing the cellulose or a cereal-based diet containing 8.4% pectin for 28 days. Nitrobenzene at 50, 100, 150, 200, 400 or 600 mg/kg of body weight was then administered. Methaemoglobin concentrations were consistently higher in rats fed diets containing pectin; at 4 h after administration of 200 mg nitrobenzene/kg of body weight, methaemoglobin levels were at background (about 6%), 31 ± 9% and 42 ± 7% in the cellulose, 5% pectin and 8.4% pectin dietary groups, respectively, while at 4 h after administration of 600 mg nitrobenzene/kg of body weight, methaemoglobin levels in the three dietary groups were 20 ± 5%, 44 ± 6% and 64 ± 1%, respectively. Caecal reductive metabolism of radioactive nitrobenzene *in vitro* was greatest in animals fed the cereal-based diet containing

8.4% pectin, followed by the purified diet with 5% pectin, then the diet without pectin.

The covalent binding of [^{14}C]nitrobenzene was investigated in erythrocytes and spleens of male B6C3F$_1$ mice and male Fischer-344 rats following single oral doses (Goldstein & Rickert, 1984). Total and covalently bound ^{14}C concentrations in erythrocytes were 6–13 times greater in rats than in mice following administration of 75, 150, 200 or 300 mg/kg of body weight, suggesting that the reported species differences in nitrobenzene-induced red blood cell toxicity may be related to the differences in erythrocytic accumulation of nitrobenzene and its metabolites. Covalently bound ^{14}C in erythrocytes peaked at 24 h in rats after administration of 200 mg/kg of body weight, whereas the low level of binding plateaued at 10 h in mice.

In the 2-week NTP gavage study described in section 7.2, reticulocyte counts were increased in male B6C3F$_1$ mice at a dose of 75 mg/kg of body weight per day, whereas methaemoglobin levels were increased in mice in all dose groups except 75 mg/kg of body weight per day males and 37.5 mg/kg of body weight per day females. Treated Fischer-344 rats showed increases in methaemoglobin and in reticulocyte counts.

In the 13-week gavage study in B6C3F$_1$ mice, described in section 7.2 (NTP, 1983a), there were increases in methaemoglobin and reticulocytes in all treated groups, most evident at the high dose (300 mg/kg of body weight per day), with decreases in haemoglobin, haematocrit and red blood cells at 150 and 300 mg/kg of body weight per day; at 75 mg/kg of body weight per day, haemoglobin was decreased. Male mice exhibited leukopenia at 18.75 and 150 mg/kg of body weight per day and leukocytosis at 300 mg/kg of body weight per day. Similarly, lymphopenia was seen in all treated males except at 300 mg/kg of body weight per day, at which dose lymphocytosis was seen. High-dose females exhibited neutrophilia and lymphocytosis. Liver and spleen haematopoiesis and splenic haemosiderin accumulation were noted in a dose-related manner at and above 75 mg/kg of body weight per day, with all animals affected at the high dose. Lymphoid depletion was noted at the two highest doses, with mainly females being affected.

In the 13-week gavage study in Fischer-344 rats, described in section 7.2, there were dose-related increases in methaemoglobin, reticulocytes, polychromasia and anisocytosis, along with decreases in haemoglobin, haematocrit and red blood cells. In the surviving high-dose animals, there was marked leukocytosis, with lymphocytosis and neutrophilia. At necropsy, the spleens of many high-dose animals were enlarged, granular and/or pitted. Nitrobenzene-treated animals had increased splenic pigment, which was usually minimal. Thickening and fibrosis of the splenic capsule were noted in all treated groups except at the lowest dose and were clearly evident at the high dose. It was considered to be an inflammatory rather than a fibrotic response, with lymphocyte, macrophage and neutrophil infiltration. There were occasional mast cells, haemosiderin-filled macrophages and fragmented necrotic cells. In a number of cases, the mesothelial cells were hypertrophied and/or hyperplastic.

In the liver, brown granular pigment was observed in the Kupffer cells of high-dose rats. The tubular epithelial cells of some high-dose rats and 75 mg/kg of body weight per day females contained pigment. The presence of pigment in the spleen, liver and kidney was considered to be secondary to the methaemoglobinaemia and anaemia and subsequent red blood cell breakdown. Accompanying this, increased haematopoiesis was seen in the bone marrow of a number of 75 and 150 mg/kg of body weight per day animals. No NOEL could be derived from this study.

Toxic haemolytic events were induced in 8-, 11- and 29-week-old Wistar WIST/RIPB rats after single oral doses of nitrobenzene at 350 mg/kg of body weight; there was a decrease in femoral bone marrow cellularity (all ages, more prominent in older rats), an increase in the erythroid:myeloid ratio in bone marrow (most evident in 11-week-olds) and large increases in rat splenic erythropoiesis (most evident in 8-week-olds) (Berger, 1990).

In the Burns et al. (1994) 2-week gavage study in B6C3F$_1$ mice, the dose was close to an MTD, with 8.5% of animals dying during the exposure period. (This study is reported in detail in section 7.8, because it was specifically designed to look at the potential immunotoxicity of nitrobenzene.) Gross histopathology revealed severe congestion of the splenic red pulp with erythrocytes and reticulocytes and haemosiderin pigmentation in the mid- and high-dose groups (100 and 300 mg/kg of body weight per day). Haematology indicated red blood cells to be the primary target, with a dose-dependent decrease in red blood cell numbers and concomitant increases in mean corpuscular haemoglobin and mean corpuscular volume and a dose-dependent increase in peripheral reticulocytes (almost 5-fold at the high dose). Leukocyte numbers were not significantly affected. In bone marrow, there were dose-dependent increases in the number of nucleated cells per femur (62% at the high dose), in DNA synthesis in the whole cell population (80% at the high dose) and in the number of monocyte/granulocyte stem cells. Increased bilirubin at the high dose could have arisen from liver damage or by increased erythropoiesis or haemolysis.

In the Mitsumori et al. (1994) repeated-dose and reproductive toxicity study in Sprague-Dawley rats, haemolytic anaemia due to methaemoglobin formation was evident in males from each treated group, with dose-related increases in erythroblasts, reticulocytes, total bilirubin and, at the high dose (100 mg/kg of body weight per day), elevations in white blood cell count.

Reactive changes secondary to haemolytic anaemia in the haematopoietic organs (haemosiderin deposition and extramedullary haematopoiesis in the liver and spleen; haematopoiesis in the bone marrow; haemosiderin deposition in the renal proximal tubular epithelium) and hepatocellular swelling were seen in all males at 60 and 100 mg/kg of body weight per day and in most males at the low dose of 20 mg/kg of body weight per day; if examined, equivalent histopathological data for females were not reported.

In the 28-day gavage study in F344 rats (Shimo et al., 1994), described in section 7.2, decreases in red blood cells, haemoglobin and haematocrit were observed in the 25 and 125 mg/kg of body weight per day groups. Histopathology revealed extramedullary haematopoiesis in the liver, brown pigmentation of renal tubular epithelium, congestion, increased brown pigmentation in red pulp, increased extramedullary haematopoiesis of the spleen and increased haematopoiesis of the bone marrow.

7.7.2   Dermal

In the 2-week NTP study in $B6C3F_1$ mice and Fischer-344 rats, described in section 7.2.2 (NTP, 1983b), reticulocyte counts and methaemoglobin levels were increased in mice and rats (all dosage groups except mice receiving lowest dose, 200 mg/kg of body weight per day); haemoglobin and red blood cells were decreased in rats.

7.7.3   Inhalation

In the CIIT 2-week study (Medinsky & Irons, 1985) in Fischer-344 rats, Sprague-Dawley (CD) rats and $B6C3F_1$ mice, dose-related increases in methaemoglobinaemia were observed in CD and Fischer-344 rats, apparent at levels as low as 51 mg/m^3 (10 ppm), reversible after 14 days of recovery. Methaemoglobinaemia was also noted in mice, although qualitative data were not reported. A marked elevation was noted in circulating white blood cells (both granulocytes and lymphocytes) in male CD rats, but not Fischer-344 rats or $B6C3F_1$ mice, exposed at 180 and 640 mg/m^3. Both rat strains, but not mice, exhibited a dose-related reversible reduction in red blood cell counts. Dose-related splenic lesions were reported in all treated animals exposed for 14 days; common findings included haemosiderosis, extramedullary haematopoiesis and sinusoidal congestion. A capsular hyperplastic lesion was seen at 180 and 640 mg/m^3 in Fischer-344 rats (of possible mesenchymal origin). In mice, a concentration-dependent increase in marginal-zone macrophages and a lymphoid hypoplasia in periarteriolar sheaths were observed (Medinsky & Irons, 1985).

In the 90-day CIIT study (Hamm, 1984; Hamm et al., 1984) in Fischer-344 rats, Sprague-Dawley (CD) rats and $B6C3F_1$ mice, methaemoglobinaemia and haemolysis, together with increases in spleen and liver weights, were observed in mice and rats at 260 mg/m^3 and in the rat strains only at 82 mg/m^3.

In the CIIT carcinogenicity study in $B6C3F_1$ mice and Fischer-344 rats, described in section 7.3, methaemoglobinaemia and anaemia were observed at and above 130 mg/m^3 in mice and rats (CIIT, 1993; Cattley et al., 1994). Methaemoglobinaemia was consistently seen at the 2-year terminal sacrifice at 130 mg/m^3 and 260 mg/m^3 in mice and at 130 mg/m^3 in rats, with an approximate 1.5- to 2-fold higher percent methaemoglobin at these exposure levels compared with controls. It was also apparent at the lowest dose of 5 mg/m^3 at the interim 15-month sacrifice of 10 rats per strain per sex per dose, with a significant increase (3–3.5 times controls in CD male rats) or a trend to an increase (up to approximately 1.5 times controls in Fischer-344 rats) in blood methaemoglobin levels — i.e., there may have been some adaptation to this effect with increased duration of exposure (Cattley et al., 1994). In both strains of rats, increases of Howell-Jolley bodies were seen. A NOEL could not be determined from these studies, as effects were observed at the lowest exposure levels studied. In mice, findings at the lowest dose of 130 mg/m^3 included reductions in haemoglobin, haematocrit and red blood cells and increases in mono-cytes and macrocytes in females. In rats, findings at the lowest dose of 5 mg/m^3 included a significant increase in methaemoglobin levels (CD rats, interim 15-month sacrifice) or a trend to an increase (Fischer-344 rats, interim sacrifice), with a trend, albeit not significant at the low dose, to changes in other haematological parameters, including an increase in platelets (female Fischer-344 rats). Histopathological findings at the low dose included a slight increase in extramedullary haematopoiesis, with splenic pigmentation and congestion.

7.8   Immunological effects

In the Burns et al. (1994) 2-week gavage study in female B6C3F$_1$ mice, described in section 7.2.1, several immunological and host resistance responses were studied.

Nitrobenzene did not change the splenic IgG response to sheep red cells, but there was a moderate suppression of the IgM antibody response on day 4 (at the two highest doses), with recovery within 20 days.

The mitogenic response (based on specific activity) of spleen cells to T-cell mitogens phytohaemagglutinin and concanavalin A was dose-relatedly suppressed, whereas no effects on the response to B-cell mitogen lipopolysaccharide were observed. The lack of effect when data were expressed as counts per minute per spleen could be explained on the basis that non-mitogen responsive cells are entering or replicating in the spleen (consistent with the observed spleno-megaly).

The mixed leukocyte response of exposed cultured spleen cells to the alloantigens was dose-dependently depressed on days 4 and 5; as with the mitogen response (above), the increase in spleen cell number can account for the suppression.

Nitrobenzene did not alter the delayed hypersensitivity response to keyhole limpet cyanin or serum complement levels.

The activity of the mononuclear phagocyte system was assessed by the ability of the reticuloendothelial system of treated mice to clear sheep red blood cells. Increased uptake of sheep red blood cells by the enlarged livers accounted for a shortened circulating half-life of these cells.

The total number of peritoneal cells was increased in high-dose mice, with a greater than proportionate increase in phagocytosis of fluorescent 0.85-μm-diameter beads.

The effect of nitrobenzene on natural killer cell activity in spleens from treated mice was assessed *ex vivo* by the lysis of YAC-1 target cells. Significant decreases in lytic capacity were seen at doses of 100 and 300 mg/kg of body weight. While the depression at the high dose could be due to a dilution of natural killer cells by non-natural killer cells (in a manner similar to that seen in the mitogen studies; see above), the 100 mg/kg of body weight group showed a marked diminution of natural killer cell activity without a reciprocal increase in the number of spleen cells.

In studies on host resistance to microbial or tumour challenge, treated mice were challenged with intravenous *Plasmodium berghei* (in mouse red blood cells), intravenous *Listeria monocytogenes*, intraperitoneal *Streptomyces pneumoniae*, intraperitoneal herpes simplex type 2 virus or intravenous metastatic pulmonary tumour, B16F10 melanoma, on day 15. Mice were not markedly more suscep-tible to *S. pneumoniae* or *P. berghei* than corn oil controls. For *S. pneumoniae*, host resistance involves complement, B lymphocytes, neutrophils and macrophages; for *P. berghei*, it involves cytophilic antibody and antibodies that enhance the capability of macrophages to recognize free plasmodia. Mortality after the herpes simplex viral challenge was reduced in nitrobenzene-treated animals compared with vehicle controls, particularly at 100 mg/kg of body weight. The modest protection may have been attributable to enhanced macrophage or interferon activity, which was able to compensate for the depressed natural killer cell and T-lymphocyte function. At the two highest nitrobenzene doses, mice were more susceptible than vehicle controls to death as a result of *L. monocytogenes* infection, especially at the 100 mg/kg of body weight dose. Resistance to this infection is mediated primarily by T lymphocytes, macrophages and complement activity. In terms of the percentage of animals with tumours, nitrobenzene treatment did not markedly affect host resistance to B16F10 melanoma, a resistance that involves T lymphocytes and macrophages, but there was a slight increase in the mean number of lung nodules at the 300 mg/kg of body weight dose (Burns et al., 1994).

In conclusion, most of the effects of nitrobenzene on the immune system in these studies can be explained by the increased cellularity of the spleen, although the new cells entering the spleen did not respond to mitogens or allogeneic cells. A degree of immunosuppression was evidenced by a diminished IgM response to sheep red blood cells, a finding that cannot be explained by an increase in spleen cells lacking immune functional capacity. Nitrobenzene at all doses stimulated the bone marrow, with increases in cells per femur, DNA synthesis and granulocyte/macrophage colony-forming unit stem cells per femur. Host resistance to microbial or viral infection was not markedly affected by nitrobenzene, although there was a trend towards increased susceptibility in cases in which T-cell function contributes to host defence (Burns et al., 1994).

No specific studies were located regarding immunological effects in animals after dermal or inhalational exposure to nitrobenzene. In toxicology studies in laboratory animals, effects on white blood cells were common, albeit with apparently inconsistent results; leukocytosis, neutrophilia and lymphocytosis were more commonly seen in rodents at high gavage doses, although leukopenia and lymphopenia were reported in other studies, generally at lower doses and more commonly by the dermal route. The increase in white blood cells may be a compensatory response to nitrobenzene-induced leukocytopenia or a response to increased infections as a result of the immunotoxicity of the compound. Details of these latter studies may be found in sections 7.2.1 and 7.2.2.

7.9   Mechanisms of toxicity

7.9.1   Methaemoglobinaemia

Nitrobenzene causes methaemoglobinaemia after all routes of exposure. The results of inhalation studies with rats and mice are compiled in Table 14. Control values for percent methaemoglobin vary between these experiments, which makes their direct comparison difficult. Nevertheless, it appears that NOELs and lowest-observed-effect levels (LOELs) are not significantly lowered by increasing the duration of exposure.

The mouse also appears to be more resistant than other species to the methaemoglobin-forming properties of nitrobenzene given by other routes of exposure (Shimkin, 1939; Smith et al., 1967); thus, even at an intraperitoneal dose of 10 mmol nitrobenzene/kg of body weight (1230 mg/kg of body weight) in CD1 female mice, which killed two-thirds of mice in 40 min, levels of methaemoglobin did not exceed 7.2% at 20 min. A similarly toxic dose of 5 mmol aniline/kg of body weight (466 mg/kg of body weight) produced no more than about 15% methaemoglobin (Smith et al., 1967). The relative resistance of mice was suggested to be due to the high activity of an NADH-dependent methaemoglobin reductase (as well as an NADPH-dependent reductase, as in cats, dogs and humans) (Stolk & Smith, 1966; Smith et al., 1967).

However, after high dermal or subcutaneous doses in C3H female mice, spectral analysis of the blood revealed methaemoglobin (Shimkin, 1939). Similarly, Stifel (1919) placed eight drops of nitrobenzene on cotton wool in an open jar containing white mice. They died within 4 or 5 h; at necropsy, the blood was almost black and showed the presence of methaemoglobin.

The action of bacteria normally present in the small intestines of rats is apparently an important element in the formation of methaemoglobin resulting from nitrobenzene exposure, since germ-free rats were reported not to develop methaemoglobinaemia when intraperitoneally dosed with nitrobenzene (Reddy et al., 1976). As noted in section 6.3, it appears that nitrobenzene metabolites formed by bacterial reduction are involved in methaemoglobin formation. After oral dosing of rats with nitrobenzene, Levin & Dent (1982) showed that the amount of

Table 14. Percent methaemoglobin in rats and mice exposed to nitrobenzene vapour[a]

Study duration	Dose (ppm)[b]	Percent methaemoglobin[c]					
		Fischer-344 rats		CD rats		B6C3F₁ mice	
		Male	Female	Male	Female	Male	Female
2 weeks[d]	0	0	3.6	6.9	4.8		
	10	1.9	4.8	6.1	6.3		
	35	6.6	6.6	8.7	7.3		
	125	11.7	13.4	14.0	31.3		
90 days[e]	0	1.2	1.6	1.4		0.7	1.3
	5	3.0*	3.2	1.6		1.6	0.8
	16	4.4*	3.9*	3.5*		2.1	2.0
	50	10.1*	10.5*	9.9*		5.8*	5.1*
15 months	0	2.9	2.4	1.2			
	1	3.2	3.33	4.1*			
	5	3.2	3.2	6.2*			
	25	4.7*	5.9*	5.9*			
24 months	0	3.9	2.7	2.8		2.0	1.4
	1	3.3	2.1	2.9			
	5	4.2	2.5	2.4		1.9	1.4
	25	5.3*	5.0*	4.6*		3.02	2.2
	50					4.0*	2.8*

Table 14 (Contd).

^a Data from Cattley et al. (1994), Medinsky & Irons (1985) and Hamm et al. (1984); results on CD rats from the 90-day study were means from animals of both sexes.

^b 1 ppm = 5.12 mg/m^3.

^c Asterisk (*) indicates significance at $P < 0.05$.

^d 5 h per day, 5 days per week; blood taken 3 days after last exposure.

^e 6 h per day, 5 days per week; sampling time not given.

the major reduced metabolite, *p*-hydroxyacetanilide, was reduced by 94% in germ-free rats.

Furthermore, correlated with an increase in *in vitro* reductive metabolism of [^{14}C]nitrobenzene by the caecal contents of rats fed purified diets containing increasing amounts of pectin, there was an increase in the ability of orally administered nitrobenzene to induce methaemoglobinaemia (Goldstein et al., 1984a).

These observations that germ-free rats apparently do not develop methaemoglobinaemia when intraperitoneally dosed with nitrobenzene and that nitrobenzene causes methaemoglobinaemia in animals (and humans) by the oral, dermal and inhalational routes seem to indicate the following possibilities: 1) there may be extensive enterohepatic recycling of absorbed nitrobenzene, regardless of the route of exposure (allowing access by nitrobenzene to the gut bacteria); 2) nitrobenzene may relatively freely transfer across membranes and access gut contents — note that even nitrobenzene vapours undergo significant absorption through the skin (see section 6.1.3); or 3) alternatively, there may be sufficient systemic metabolism of nitrobenzene to generate haemoglobin-reactive species. In this regard, the work of Levin & Dent (1982) and others indicates that in rats there is some capacity for systemic oxidative and reductive metabolism, even if gut bacterial reduction is the primary *in vivo* step.

Apart from the work of Rickert et al. (1983) utilizing bile duct cannulation following oral dosing, there do not appear to be any bile duct cannulation experiments specifically designed to look at the issue of enterohepatic cycling of nitrobenzene after other routes of adminis-tration. Rickert et al.'s (1983) data indicated that, after oral dosing at least, the extent of enterohepatic cycling was limited. They suggested that faecal radioactivity and reduced metabolites (after both oral and intraperitoneal dosing in Fischer-344 rats) entered the gut during the distribution phase after administration rather than by biliary excretion or incomplete absorption after oral administration. (This suggestion would support the second possibility outlined in the previous para-graph.)

It appears that most extensive metabolic studies that have been conducted on nitrobenzene (which have been published) have utilized oral dosing, so it is not possible to make a back-to-back comparison of quantitative and qualitative differences in nitrobenzene metabolism depending on the absorption route. Rickert et al. (1983) noted little difference in the pattern of excretion of total radioactivity after oral and intravenous dosing in CD rats, but there did not appear to be data on comparative metabolism.

Aniline originally was considered responsible for methaemoglobin formation following nitrobenzene exposure. Although all is still not completely understood, it appears that the reactive intermediates in the nitrobenzene reduction pathway, nitrosobenzene and phenylhydroxylamine, may be involved in oxidation–reduction cycling with haemoglobin. Both intermediates can produce methaemoglobin if either or both are injected *in vivo* (Kiese, 1966). It has also been suggested that oxidative damage to red blood cells may arise from hydrogen peroxide formed as a result of "auto-oxidation" of quinone intermediates such as *p*-aminophenol. This is based on the fact that *p*-aminophenol itself, given *in vivo*, produces methaemoglobin (Kiese, 1966). In addition, superoxide free radicals may be generated in a futile reaction cycle during nitrobenzene metabolism (see Figure 5 in chapter 6). The parent nitro compound is regenerated in this redox cycle, with the only new products cycling being expended $NADP^+$ and superoxide anion radical (Levin & Dent, 1982). In relation to the damage that superoxide anions may cause, it is known that superoxide dismutase is, in addition to methaemoglobin reductase, an important enzyme in the oxidant protection of erythrocytes (Luke & Betton, 1987).

Catalase, the enzyme that catalyses the decomposition of hydrogen peroxide, is inhibited in red blood cells following the absorption of nitrobenzene (Goldstein & Popovici, 1960) and has been reported to be inhibited by very low concentrations of the nitrobenzene metabolites hydroxylaminobenzene and *p*-aminophenol (De Bruin, 1976). With respect to scavenging of peroxides in biological systems, there are several glutathione peroxidases that can scavenge hydrogen peroxide and organic peroxides, substrates for which these enzymes have high affinity. Catalase, on the other hand, can scavenge only hydrogen peroxide, for which it has low affinity. Nevertheless, in an *in vitro* test system, a 20–200 times faster rate of hydrogen peroxide formation was needed to produce methaemoglobin in normal red blood cell suspensions than in suspensions of red blood cells lacking catalase (De Bruin, 1976); thus, inhibition of catalase in red blood cells by nitrobenzene may contribute to its potency in causing met-haemoglobin formation.

Data from Goldstein & Rickert (1985) showed that nitrobenzene did not increase methaemoglobin formation when incubated with red blood cell suspensions prepared from rats (Fischer-344 males). This was not due to lack of transfer across red blood cell membranes, since radioactive nitrobenzene accumulated to the same extent as or to a greater extent than *o-, m-* and *p-*dinitrobenzene, compounds that did cause increased methaemoglobin. This process was simple partitioning rather than active transport, since "uptake" was not affected by temperature and was maximal within 1 min. Furthermore, there was no methaemoglobin formation when nitrobenzene was incubated with haemolysates.

Results indicated that aniline, hydroxylaminobenzene, nitroso-benzene and nitrobenzene are all metabolized *in vivo* to yield the same metabolite, most probably phenylhydronitroxide radicals (produced from the reaction of hydroxylaminobenzene with oxyhaemoglobin), which are responsible for the oxidation of thiols within red blood cells (Maples et al., 1990).

Anaemia is caused by methaemoglobin formation, by altered globin chains at sites of thiol-containing amino acids, such as cysteine, and by red blood cell lysis.

7.9.2   Splenic toxicity

Splenic toxicity is related to erythrocyte toxicity, since a primary function of the spleen is to scavenge senescent or damaged red blood cells. Splenic capsular lesions were seen in rats and mice (e.g., Kligerman et al., 1983; Hamm et al., 1984); splenic engorgement and increased covalent binding of nitrobenzene in the spleen result from the haematotoxicity of the compound and the resultant scavenging of damaged erythrocytes (e.g., NTP, 1983a, 1983b; Goldstein & Rickert, 1984). This is supported by the observation that splenic engorgement is less apparent in mice than in rats, correlated with differences in species susceptibility to nitrobenzene-induced red blood cell damage (Goldstein et al., 1983b). The mechanism that causes splenic capsular thickening may relate to this scavenging and compensatory splenic haematopoiesis.

Splenic injury may arise from the deposition of massive amounts of iron or other red blood cell breakdown products in the spleen, or reactive metabolites of nitrobenzene might be delivered to the spleen, their subsequent reaction with splenic macromolecules causing organ toxicity. Yet another feasible mechanism is the accumulation of red blood cell enzymes, which could produce reactive intermediates from nitrobenzene already in splenic tissue (Bus, 1983). It is unlikely that splenic damage will occur without erythrocyte damage.

Similar splenic lesions have previously been observed with aniline and aniline-based dyes, some of which produced splenic sarcomas in chronic carcinogenicity studies in rats (Hazleton Laboratories, 1982; see also section 7.10).

7.9.3 Renal toxicity

Differences in species and possibly strain susceptibility to the renal effects of nitrobenzene exposure are apparent in toxicology. Observed effects in laboratory animals have included increased kidney weights, pigmentation of tubular epithelial cells, hydropic degeneration of the cortical tubules and hyaline nephrosis, and swelling of the glomeruli and tubular epithelium. In short-term vapour inhalation studies in CD and Fischer-344 rats, hydropic degeneration of the cortical tubular cells was observed in CD rats (predominantly in females), and hyaline nephrosis was seen only in Fischer-344 rats (predominantly in males). By the same route, renal effects in $B6C3F_1$ mice included degenerative changes in tubular epithelium of males, but neither hydropic degeneration of the cortical tubular cells nor hyaline nephrosis was seen (see section 7.2.3).

It is possible that the hyaline nephrosis seen in male rats is due to a mechanism involving alpha$_{2u}$-globulin. A number of chemicals, including unleaded gasoline, 2,2,4-trimethylpentane, 1,4-dichlorobenzene and *d*-limonene, a natural product found in citrus oils, have been found to cause kidney tumours specific to male rats by binding to a specific protein in the proximal tubules of male rats, alpha$_{2u}$-globulin, leading to hyaline droplet accumulation (Dietrich & Swenberg, 1991; Hard et al., 1993; Borghoff, 1999). This protein is not found in female rats, mice or humans. It has been concluded (e.g., US EPA, 1991; IARC, 1999) that renal pathology and tumours involving alpha$_{2u}$-globulin are specific to male rats and do not contribute to the weight of evidence that a chemical poses a carcinogenic hazard in humans. However, hyaline droplet accumulation can also be a response to overload of other proteins in the renal tubule, and adequate characterization of kidney pathology is needed to help differentiate alpha$_{2u}$-globulin inducers (which are not relevant to human cancer risk assessment) from chemicals that may produce renal pathology and, possibly, tumours through other means.

At this stage, it is not possible to conclude that nitrobenzene causes nephropathy by an alpha$_{2u}$-globulin mechanism, in view of the fact that the available toxicology studies, including the CIIT carcinogenicity study (study started in 1983; see section 7.3), did not report on all the criteria needed to distinguish this specific mechanism (US EPA, 1991; IARC, 1999). Another factor that makes it difficult to come to a conclusion as to whether nitrobenzene acts by this mechanism is the lack of concordance in the kidney findings in male rats of the F344 and Sprague-Dawley (CD) strains. In the chronic inhalation study (section 7.3), eosinophilic droplets noted as spherical intracytoplasmic inclusions within the proximal convoluted tubules showed a concentration-related incidence in male F344 rats (and were also elevated in the high-dose female group), whereas similar findings were not reported in male Sprague-Dawley (CD) rats. Similarly, whereas nitrobenzene at the high dose (130 mg/m^3) caused kidney tubular hyperplasia and neoplasias (adenomas and carcinomas) only in male, and not female, Fischer F344 rats, neoplasias were not seen in male Sprague-Dawley (CD) rats; there may have been a marginal increase in the incidence of tubular hyperplasia at the high dose (incidence of 5%, 1%, 7% and 9% at 0, 5, 26 and 130 mg/m^3, respectively). For those chemicals that have been shown to cause kidney tumours by an alpha$_{2u}$-globulin-dependent mechanism, it is noted that males of both Fischer-344 and Sprague-Dawley strains show alpha$_{2u}$-globulin accumulation and protein droplet nephropathy in response (IARC, 1999). On the other hand, the finding that kidneys of nitrobenzene-exposed B6C3F$_1$ mice showed neither eosinophilic droplets nor tubular hyperplasia and neoplasia is consistent with the proposed alpha$_{2u}$-globulin mechanism.

7.9.4 *Neurotoxicity*

In acute (section 7.1) and subchronic studies in rodents (section 7.2), lesions in the brain stem and cerebellum were the most life-threatening toxic effects seen. In severe methaemoglobinaemia arising from extensive nitrobenzene poisoning, central nervous system effects may be predicted on the basis of hypoxia alone. It has also been hypothesized that these lesions might represent a hepatic encephalopathy secondary to the liver toxicity of nitrobenzene (Bond et al., 1981). Other results suggest that it is possible that brain parenchymal damage may have resulted from anoxia or hypoxia due to vascular damage or decreased blood flow to affected areas (see section 7.2.2).

Another possible mechanism for the central nervous system damage is the formation of superoxide radicals or toxic hydroxyl radicals generated from hydrogen peroxide (see discussion in section 7.9.1). Evidence has been adduced to indicate that the ability of a related compound, dinitrobenzene, to cause cell death in *in vitro* co-cultures of rat brain astrocytes and brain capillary endothelial cells (a blood–brain barrier model) is at least partly due to the generation of hydroxyl radicals in the culture (Romero et al., 1996).

7.9.5   *Carcinogenicity*

Nitrobenzene is carcinogenic in experimental animals, but apparently not via a genotoxic mechanism. Several other mechanistic explanations have been put forward.

1)   Oxidation mechanisms

Nitrobenzene is oxidized by various ring microsomal oxygenases to oxygenated ring forms, including aminophenolic and nitrophenolic compounds (see Figure 3 in chapter 6) (Robinson et al., 1951; Parke, 1956; Rickert et al., 1983). Oxidation produces mostly *p*-nitrophenol, *p*-aminophenol and *p*-hydroxyacetanilide metabolites in the excreta (Parke, 1956; Ikeda & Kita, 1964; Rickert et al., 1983). Certain metabolites are made more polar by metabolic sulfation, acetylation or glucuronidation.

In tissues with sufficient oxygen, the nitroanion free radical can be oxidized by oxygen in a "futile reaction," generating pernicious amounts of tissue superoxide anions while regenerating parent nitrobenzene (see Figure 4B in chapter 6) (Mason & Holtzman, 1975b; Sealy et al., 1978; Bus & Gibson, 1982; Levin et al., 1982; Bus, 1983). This futile reaction may account for a number of the toxic, carcinogenic actions of nitrobenzene, based on a sustainable pool of persistent nitroxide intermediates and the known carcinogenic properties of the superoxide anion radical (Flohé et al., 1985; Trush & Kensler, 1991; Guyton & Kensler, 1993; Cerutti, 1994; Feig et al., 1994; Dreher & Junod, 1996).

The elimination kinetics in urine of the nitrobenzene metabolite *p*-nitrophenol are slow, suggesting that nitrobenzene is either recycled in the bile or retained by other means (Salmowa et al., 1963; Rickert, 1984). There is experimental evidence that bile recycling may not be significant, and Rickert (1987) suggested that the retaining action may be the "oxidation futile reaction," which may continually regenerate nitrobenzene, thereby slowing its net elimination from the body. This slow elimination characteristic of nitrobenzene may account for some of its toxicity. The carcinogenic effects would increase with increased tissue concentrations and the increased residence time of nitrobenzene (and its metabolites).

2)   Reduction mechanisms

In the caecum, endogenous bacteria efficiently convert orally ingested nitrobenzene to reduced nitroxide intermediates. This reduction is mechanistically a concerted two-electron per step process from nitrobenzene to nitrosobenzene to phenylhydroxylamine to aniline (see Figure 4A in chapter 6; Holder, 1999a). Whereas oral exposure results in the formation of nitroxides in the caecum by bacterial nitro-reductases, inhalation exposure produces nitroxides in cellular micro-somes (and possibly the mitochondria) by different nitroreductase enzymes (Wheeler et al., 1975; Peterson et al., 1979; Levin & Dent, 1982). Once nitrobenzene is orally absorbed, the microsomal one-electron per step reduction process produces reduced nitroxides, with aniline being the final product of that reduction sequence (see Figure 4B in chapter 6). By the inhalation route, the enteral reduction process should be largely bypassed, and hence system microsomal reduction would be expected to be the predominant reduction mechanism.

While the chemically reactive intermediates nitrosobenzene and phenylhydroxylamine are produced in both reduction processes, only the one electron per step reduction also produces associated free radical intermediates — e.g., the nitroanion free radical (see Figure 4B in chapter 6) (Mason & Holtzman, 1975a; Mason, 1982). Whereas only modest steady-state levels of nitrosobenzene and phenylhydroxylamine actually occur in rat liver, as directly measured by ESR, circulating red blood cells have significant and stable specific ESR signals, indicating persistent amounts of these two nitroxides (Eyer et al., 1980; Blaauboer & Van Holsteijn, 1983). Nitrosobenzene and phenylhydroxylamine drive reactions forming methaemoglobin and consuming NAD(P)H, thus maintaining a persistent redox couple, nitrosobenzene and phenylhydroxylamine (see Figure 5 in chapter 6) (Eyer & Lierheimer, 1980). Hence, it may be concluded that frequent nitrobenzene re-exposures, as in the chronic rodent bioassay, tend to initiate and maintain the cycling actions of the redox couple, nitrosobenzene and phenylhydroxylamine. This redox maintenance could also occur in industrial exposures. The redox couple in red blood cells constitutes an ongoing catalytic pool that resists nitrobenzene metabolic clearance and could affect many tissue types. This couple is likely to contribute to the slow kinetic elimination of nitrobenzene, in addition to the futile reaction proposed by Rickert (1987). Because the circulation involves all tissues, the redox couple — and its driving electronic action via free radical generation — probably accounts in part for the pervasive and stable system toxicity set up by nitrobenzene exposure. This pervasiveness may explain why each of the three species in the CIIT bioassay responded with tumours at eight organ sites, at least at the highest exposure levels tested. By extension, humans would also generate the redox couple and therefore would be likely to respond with cancer if exposed in a manner analogous to the rodents.

Nitrosobenzene can bind glutathione to form a relatively stable circulating glutathione–nitrosobenzene conjugate. Figure 5 in chapter 6 suggests that this conjugate can translocate throughout the body where it can 1) homeolytically cleave to form the reactive glutathiyl radical; 2) undergo redox to form phenylhydroxylamine; or 3) rearrange to form glutathione sulfinamide, which in turn cleaves to produce aniline (Eyer, 1979; Eyer & Lierheimer, 1980; Eyer & Ascherl, 1987; Maples et al., 1990).

3) Other metabolic considerations

Because of the ubiquity of the redox conditions capable of producing aminophenols, nitrophenols, nitrosobenzene and phenylhydroxylamine in various organs, a variety of tissues can be damaged.

Because of translocation and free radical chain reactions, this may not necessarily be confined to tissues where the metabolites or free radicals were originally produced. Specific toxicity profiles in different organs depend on detoxifying enzyme levels and many host- and tissue-specific factors, such as the number of endogenous free radical producers, quenching agents, spin traps (agents acting as stabilizers) and carriers (agents acting to transport free radicals) (Stier et al., 1980; Keher, 1993; Gutteridge, 1995; Netke et al., 1997).

Any aniline produced from nitrobenzene may serve as a pool to be later oxidized to reform the nitroxide intermediates, which would further reinforce the redox couple and resist nitrobenzene metabolite clearance, hence contributing to the slow elimination. Although aniline is the final product of both reduction sequences, it is likely that the nitroxide intermediates nitrosobenzene and phenylhydroxylamine and their associated free radicals in the one electron per step process (see Figure 4B in chapter 6) are the most chemically reactive and hence the most likely cause of toxicity. As an indication of aniline's reservoir activity, aniline's oxidation has been linked to lipid peroxidation (Stier et al., 1980; Khan et al., 1997).

Critical redox imbalances are likely to occur in various cells in rodents and humans, and these imbalances are, in part, clinically manifest by haemosiderosis, methaemoglobinaemia, anaemia, testicular atrophy and liver, spleen and brain effects at sufficiently high exposures. At such exposures, nitrobenzene reduction and oxidation processes are likely to cause redox imbalances — at least for some tissues. Hence, nitrobenzene is likely to be toxic in a "context-sensitive manner." That is, where free radicals are poorly quenched or trapped and/or oxygenated ring products are not conjugated and efficiently eliminated, then there exist conditions that may lead to chemical carcinogenesis. Based on sustained redox imbalances, there may be thresholds for carcinogenicity in some tissues, while other tissues may not exhibit a practical threshold. At the moment, bioassay design and the understanding of the toxicokinetics of nitrobenzene preclude the determination of whether a low-dose threshold exists for the effects of nitrobenzene in any tissue.

Nitrobenzene has structure–activity relationships with other aromatic nitro and amino compounds that produce common reactive nitroxide intermediates — aromatic nitroso and hydroxylamine compounds and their free radicals. These similarities relate to their mutagenicity and metabolic imbalances, which can lead to cancer (Kiese, 1966; Miller, 1970; Weisberger & Weisberger, 1973; Mason, 1982; Blaauboer & Van Holsteijn, 1983; Rickert, 1987; Rosenkranz, 1996; Verna et al., 1996). Nitroaromatics are of concern as chemical carcinogens because of their metabolic activation in various environmental media (Miller, 1970; Rosenkranz & Mermelstein, 1983; Rickert, 1984) and occurrence in complex mixtures, such as municipal waste incineration emissions, diesel emissions, azo dyes and food pyrolysates (King et al., 1988; Crebelli et al., 1995; DeMarini et al., 1996). Other carcinogenic nitroxide examples are the tobacco products 4-(methylnitrosoamino)-1-(3-pyridyl)-1-butanone and *N'*-nitrosonornicotine, which are microsomally activated intermediates of tobacco combustion (redox) ingredients, nicotine and related plant alkaloids (Hecht et al., 1994; Staretz et al., 1997). These nitroxides are linked to human lung, oral cavity, oesophagus and pancreas cancers arising from direct and/or indirect sources such as passive smoking (Hecht, 1996; Pryor, 1997). Free radicals are currently being analysed in cigarette smoke, the toxicological activity of which is noted to be dependent on host factors such as vitamin concentrations, dietary lipids, and superoxide dismutase, catalase and cytochrome P-450 activities (Kodama et al., 1997; Maser, 1997). It is likely that nitrobenzene carcinogenicity is also dependent on these host factors. Whether humans resolve the free radicals better than, the same as or less efficiently than rodents remains to be demonstrated.

The NTP has bioassayed 16 nitroarenes for carcinogenicity, and 62.5% (10/16) of them are positive in mouse and/or rat bioassays. Further structure–activity relationship analysis of NTP data suggests that other functional groups, such as multiple strong electron-withdrawing groups, can suppress nitroarene carcinogenicity. Therefore, the mere presence of a nitro group in a compound does not necessarily connote carcinogenicity (Rosenkranz & Mermelstein, 1983). A systematic study has been done on certain amines that might produce nitroxide intermediates. Eleven of the chosen amines were known to be carcinogens, and eight were not carcinogenic (Stier et al., 1980). These authors experimentally found the characteristic nitroxide ESR signal (stable free radicals) in 91% of the carcinogenic amines, but only 25% of the non-carcinogenic amines generated this signal. This study suggests that those amines that generate the nitroxide ESR signal in metabolism have a tendency to be involved with carcinogenesis. No analogous systematic study was located for nitro-arenes.

7.10 Toxicity of metabolites and interactions

7.10.1 Nitrophenol

A dose-dependent increase in the formation of methaemoglobin was seen in cats after oral exposure to *o*-nitrophenol and in rats after exposure by inhalation to *p*-nitrophenol. After repeated exposure to *p*-nitrophenol by inhalation, the formation of methaemoglobin was shown to be the most critical end-point, and it is assumed to be a relevant end-point for oral exposure too (IPCS, 2000).

In mice, the dermal application of 4-nitrophenol for 78 weeks gave no indication of carcinogenic effects. In another mouse study, which had several limitations, no skin tumours were noted after dermal application of *o*- or *p*-nitrophenol over 12 weeks (IPCS, 2000).

For *p*-nitrophenol, the available data gave no evidence of repro-ductive or developmental toxicity effects after dermal or oral appli-cation to rats and mice. In an oral study with rats, *o*-nitrophenol induced developmental effects in the offspring only at doses that also produced maternal toxicity (IPCS, 2000).

7.10.2 Aniline

Aniline is formed in significant amounts from nitrobenzene by bacterial reduction in the intestines of animals and humans.

Mice given large doses of aniline showed signs of central nervous system toxicity, but only relatively limited methaemoglobinaemia (see section 7.9.1).

Aniline hydrochloride was tested for carcinogenicity in experi-ments in mice and rats by oral administration. No increase in tumour incidence was observed in mice. In rats, it produced fibrosarcomas, sarcomas and haemangiosarcomas of the spleen and peritoneal cavity (IARC, 1982). In several limited studies, largely negative results were obtained following oral administration to rats (IARC, 1982), after subcutaneous injection in mice (IARC, 1982) and hamsters (Hecht et al., 1983) and after single intraperitoneal injection of mice (Delclos et al., 1984).

There was no evidence of embryolethal or teratogenic effects observed in the offspring of rats administered aniline hydrochloride during gestation. Signs of maternal toxicity included methaemoglobinaemia, increased relative spleen weight, decreased red blood cell count and haematological changes indicative of increased haematopoietic activity (Price et al., 1985).

7.10.3 *Interactions with other chemicals*

Synergism between orally administered nitrobenzene and six other common industrial compounds (formalin, butylether, aniline, dioxane, acetone and carbon tetrachloride) was demonstrated in rat studies using death as the end-point (Smyth et al., 1969).

When alcohol was given orally and nitrobenzene was given intravenously, there was increased toxicity in rabbits. Alcohol also enhanced the neural toxicity of nitrobenzene in rabbits when nitrobenzene was applied to the skin (Matsumaru & Yoshida, 1959).

8. EFFECTS ON HUMANS

8.1 Poisoning incidents

8.1.1 Oral exposure

There are a very large number of clinical reports of poisonings arising from ingestion of nitrobenzene (Dodd, 1891; Hogarth, 1912; Zuccola, 1919; Scott & Hanzlik, 1920; Wright-Smith, 1929; Fullerton, 1930; Leader, 1932; Carter, 1936; Wirtschafter & Wolpaw, 1944; Chambers & O'Neill, 1945; Chapman & Fox, 1945; Nabarro, 1948; Parkes & Neill, 1953; Myślak et al., 1971; Harrison, 1977; Schimelman et al., 1978). In a literature review that was by no means comprehensive, Von Oettingen (1941) listed 44 published case reports that appeared between 1862 and 1936. One reason for this high incidence of poisonings arose from the fact that, following its discovery, nitrobenzene was thought to be a cheap substitute for oil of bitter almonds and was used in perfume (e.g., soaps) and in sweets; in many countries, it was handled freely under the name "oil of mirbane."

The initial symptoms of nitrobenzene poisoning, which may be delayed for up to 12 h after ingesting the poison, are those of gastric irritation (e.g., nausea and vomiting). Methaemoglobinaemia, indicated by cyanosis, which is normally unaccompanied by respiratory distress, usually develops in a few hours, but in some cases it may be apparent within an hour. In severe poisoning, usually suicidal, but rarely accidental, neurological symptoms, namely progressive drowsiness and coma, lead to death from respiratory failure, as in the case discussed by Chambers & O'Neill (1945). These symptoms are ascribed to cerebral anoxia as well as a possible direct action of nitrobenzene on the nervous system. In severe poisoning, after ingestion but not, as a rule, after inhalation, haemolytic anaemia may also occur on or about the fifth day (Nabarro, 1948).

Nitrobenzene has been ingested as an abortifacient in doses of 15–100 g up to 400 g. In a review by Spinner (1917), 7 of 16 women died and only 1 aborted; 1 recovered after ingesting, it was claimed, 400 g of the poison. In a later review by Von Oettingen (1941), many more literature reports of fatalities arising from nitrobenzene's use as a presumed abortifacient were cited.

Several relatively recent reports of poisoning by the oral route are detailed below.

A 21-year-old man was thought to have taken about 30–40 ml of a nitrobenzene-containing dye used in screen printing about 30 min before admission to hospital. He was reported to have peripheral and central cyanosis; pupils were normal size, heartbeat was 160 beats per minute, blood pressure was 80/54 mmHg and respiration was 28 per minute. Blood samples were dark brown. After 1 h of positive-pressure ventilation, gastric lavage and intravenous fluids, the patient became conscious and well oriented, with a decrease in heart rate and an increase in blood pressure. Serum methaemoglobin was 4.29 g/dl. A slow intravenous infusion of ascorbic acid was started, and methylene blue was injected intravenously; after 35 min, the colour of the patient changed dramatically from brownish-blue to pink. After a second injection of methylene blue and a transfusion of packed red blood cells, methaemoglobin was 0.6 g/dl. A peripheral blood smear revealed evidence of haemolytic anaemia, but there was no evidence of occult blood in the urine. The patient was discharged on the fifth day of admission (Kumar et al., 1990).

A 33-year-old man injected 4 ml of India ink into a median cubital vein with suicidal intent. He was hospitalized in good general condition 10 h after the injection. Abnormal laboratory test results were leukocytosis, a methaemoglobin level of 36.9% (normal range 1.5%) and a free haemoglobin level of 74 nmol/litre (normal range <25 nmol/litre). The presence of nitrobenzene in blood and urine was demonstrated. Intravenous administration of vitamin C and tolonium chloride plus forced diuresis led to an improvement in cyanosis and a fall in the methaemoglobin concentration. Repeated increases in the concentration of aminobenzene were successfully treated by haemodialysis (Ewert et al., 1998).

A review of published reports does not provide any consistent indication of the likely acute toxicity of nitrobenzene in humans. Some reports reported recovery after massive doses of nitrobenzene in amounts of as much as 400 g, whereas other studies indicated that nitrobenzene is much more toxic (Dodd, 1891; Spinner, 1917;

Walterskirchen, 1939; Wirtshafter & Wolpaw, 1944; Moeschlin, 1965). Von Oettingen (1941) cited results of incidents in which oral intakes of about 4 ml caused deaths, 8–15 drops caused toxic effects and 20 drops were fatal. In a fairly detailed clinical report given by Moeschlin (1965), a 23-year-old florist swallowed 1–2 ml of nitrobenzene (suspended in a little water) to procure an abortion and was at "the point of death" when admitted to hospital 3 h later. Polson & Tattersall (1969) claimed that even 12 ml reported by Parkes & Neill (1953) was probably on the high side; they claimed that a toxic dose of the order of 15 drops, as reported by Smith & Fiddes (1955), was likely. The lethal dose has been claimed to be as little as 1 ml (Thienes & Haley, 1955). A more recent detailed report more reliably indicated that about 5–10 ml of nitrobenzene (near 200 mg/kg of body weight) could be fatal in the absence of medical intervention (Myślak et al., 1971). Myślak et al. (1971) estimated that, following ingestion of about 4.3–11 g of nitrobenzene, the initial level of methaemoglobin was 82% at about 90 min after ingestion, accompanied by severe symptoms, including unconciousness, cyanosis, circulatory insufficiency and rapid and shallow breathing.

8.1.2   Dermal exposure

Cases of severe and nearly lethal toxic effects after dermal exposure to aniline-based dyes were reported as early as 1886; the incident involved 10 infants in a maternity ward exposed to dye-stamped diapers (Gosselin et al., 1984). Between 1886 and 1959, there were numerous further reports of babies with a cyanotic syndrome arising from the use of laundered hospital diapers freshly stamped with an identifying mark using ink dissolved in aniline or nitrobenzene; the most recent published observation of an outbreak of methaemoglobinaemia in a number of babies was by Ramsay & Harvey (1959).

The other common dermal poisonings arose in persons wearing freshly dyed shoes. The resulting condition was often termed "nitrobenzene poisoning," even though exposures may have been to nitrobenzene or aniline (Von Oettingen, 1941; Gosselin et al., 1984). A poisoning episode involving 17 men at a US army camp was traced to the wearing of newly dyed shoes or puttees; the shoe dye contained nitrobenzene (Stifel, 1919). The soldiers experienced headache, nausea, dizziness and general malaise and were cyanosed. Cyanosis and ill-effects of nitrobenzene, including methaemoglobinaemia, were also demonstrated when two volunteer soldiers wore boots that had been saturated with a nitrobenzene-containing dye for 6 h.

Nine cases of poisoning by shoe dye were reported by Muehlberger (1925). Here, also, cyanosis of sudden onset was the outstanding feature. The poisoning was ascribed to aniline solvent in six cases, nitrobenzene in the remainder.

Poisonings including fatalities have also occurred when nitrobenzene was spilt on clothing (Stifel, 1919); most such reports indicate that it is absorbed with great rapidity and can rapidly cause severe intoxication (Moeschlin, 1965).

8.1.3 Inhalation exposure

No reports of fatalities after exposure to nitrobenzene by inhalation were retrieved, but poisoning by inhalation can be severe (Simpson, 1965). Children may be particularly susceptible to poisoning by inhalation.

Twins aged 3 weeks slept on a mattress that had been treated with a disinfectant intended to kill bed bugs. The fluid had been applied only to the edges of the mattress but was sufficient to produce an odour that had given the mother an intense headache half an hour before putting the babies to bed for 11 h. The bodies of the infants were not in contact with the poison. The next morning both had lead-coloured fingernails and toenails and were of ashen hue. It was subsequently shown that the disinfectant contained nitrobenzene. Blood analysis showed the presence of methaemoglobin. They recovered after about 8 days (Stevens, 1928).

Zeligs (1929) described the case of a 2-month-old child who was put to sleep in a crib several feet away from the mother's bed, the mattress of which had been treated with nitrobenzene for bed bugs. Within an hour, the child became blue and cried all night. Next morning, the cyanosis had deepened and she vomited. There was complete recovery after 4 days.

Sanders (1920) reported poisoning of a man (marked cyanosis) who wore newly dyed shoes when visiting a theatre for the afternoon, where he sat in an ill-ventilated place; the report did not rule out direct dermal absorption of nitrobenzene, in addition to the inhalation exposure. Several other anecdotal reports even include ill-effects from the vapours arising from the use of "almond"-glycerine soap in a warm bath.

8.2 Occupational exposure

Von Oettingen (1941) listed 14 reports published between 1862 and 1939 regarding industrial poisonings, stating that there were many other reports as well. More than 50% of the workers of a nitrobenzene plant reported spells of cyanosis, 20% showed pallor and subicteric coloration, 33% suffered from abnormal congestion of the pharynx and 45% experienced intermittent headache (Heim de Balsac et al., 1930).

Severe methaemoglobinaemia was reported in a 47-year-old woman who was occupationally exposed to nitrobenzene at unmeasured levels for 17 months (Ikeda & Kita, 1964). She was involved in painting lids of pans with a paint that contained nitrobenzene as a solvent (99.7% of the distillate from the paint). Since the poisoning arose after the workshop was remodelled and the ventilation became rather poor, it appears that the main route of exposure was via inhalation. Hepatic effects were evidenced by the fact that the liver was enlarged and tender and the results of liver function tests were abnormal. The spleen was enlarged and tender. Neurological effects included headache, nausea, vertigo, confusion and hyperalgesia to pinprick. Absorption of nitrobenzene was shown by the detection of p-nitrophenol and p-aminophenol in the urine.

Rejsek (1947) briefly reported that cyanotic workers, employed in factories producing aniline and nitrobenzene, became more cyanotic and/or fainted if exposed to bright sunlight. No reason for this observation was given.

Flury & Zernik (1931) compiled data on the toxic symptoms produced in humans by various concentrations of nitrobenzene vapour in air (Table 15).

Pacséri et al. (1958) investigated possible correlations between exposure and clinical signs and/or clinical biochemistry changes in workers in a factory in which nitrobenzene and dinitrochlorobenzene

Table 15. Toxicity of nitrobenzene vapours for humans

Reported findings[a]	Concentration in air	
	mg/litre	ppm[b]
Slight symptoms after several hours' exposure	0.2–0.4	40–80
Maximum amount that can be inhaled for 1 h without serious disturbances	1.0	200
Tolerated for 6 h without material symptoms	0.3–0.5	60–100
Tolerated for 0.5–1 h without immediate or late effects	1.0–1.5	200–300

[a] Compiled by Flury & Zernik (1931).
[b] Approximate values only. 1 ppm = 5.12 mg/m³.

were produced; the range of daily average air concentrations of nitrobenzene was 15–29 mg/m³ (mean 20 mg/m³). No anaemia was seen, but it was stated that workers showed increased methaemoglobin levels and the formation of Heinz bodies. In workers exposed to nitrobenzene and related aromatic nitro compounds, a mean value for methaemoglobin was 0.61 g/100 ml; with the method used, the upper reference limit for non-exposed people was 0.5 g/100 ml. For 2.8% of the workers, Heinz body formation was above 1%. Concentrations of nitrobenzene as high as 196 mg/m³ had been measured in 1952 in the same plant. At this earlier time, cases of anaemia and "intoxication" of workers had been reported.

Harmer et al. (1989) studied eight process operators in an anthraquinone plant in the United Kingdom. The workers operated on a 12-h shift system, with 3 days on and 3 days off. Blood was sampled on "pre-shift day 1 and post-shift day 3" for nitrobenzene and methaemoglobin measurement. Urine samples were collected pre-shift on day 1, at end-of-shift on days 1, 2 and 3, on awakening on day 4 (i.e., first rest day) and on return to work, pre-shift on day 1; *o*-, *m*- and *p*-nitrophenol were measured by HPLC. Atmospheric nitrobenzene levels over an 8-h period were measured by GC and ranged from about 0.7 to 2.2 mg/m³. Small amounts of unchanged nitrobenzene were detected in blood at pre-shift on day 1 (ranging from 0 to 52 μg/litre) and post-shift on day 3 (ranging from 20 to 110 μg/litre), indicating some accumulation of nitrobenzene in the body. Methaemoglobin levels were all below 2%, with no clear correlation with blood nitrobenzene levels. Urinary *p*-nitrophenol tended to increase over the 3-

day shift period (ranged between about 0.2 and 5.4 mg/litre), although there did not appear to be much correlation with atmospheric concentrations of nitrobenzene. Urinary *m*- and *o*-nitrophenols were "detected." No toxic signs or symptoms were reported.

8.3   Haematological effects

When nitrobenzene is ingested, the outstanding systemic effect is methaemoglobin formation. A latency period (after ingestion and before any signs or symptoms occur) can be as short as 30 min or as long as 12 h. Usually, the higher the dose, the shorter the latency period (WHO, 1986). Similarly, methaemoglobinaemia is the predominant effect of dermal and inhalational absorption of nitrobenzene.

Methaemoglobin is not capable of binding oxygen for normal gas transport. This hypoxia is generally associated with fatigue, weakness, dyspnoea, headache and dizziness. A given level of methaemoglobin produces a more severe impairment of peripheral oxygen transport than an equivalent true anaemia, because a smaller proportion of blood oxygen is released from the residual oxyhaemoglobin in tissue capillaries (Gosselin et al., 1984). Even under normal conditions, some methaemoglobin is formed (1–4%) as blood is oxygenated in the lungs (Fischbach, 1996).

At 15–20% methaemoglobin in the blood, a distinct cyanosis or slate-blue coloration is noted, whereas at 30–50% levels, the patient becomes symptomatic, with lethargy, vertigo, headache and weakness; there is moderate depression of the cardiovascular and central nervous systems, manifest as stupor, tachycardia, hypotension and respiratory depression. At greater than 60%, stupor and respiratory depression occur, which require immediate treatment (Schimelman et al., 1978).

In normal individuals, methaemoglobin is reduced by NADH-dependent methaemoglobin reductase, but two other pathways are involved: 1) glutathione reductase reduces oxidized glutathione to glutathione, which removes oxidants capable of reacting with haemoglobin to produce methaemoglobin; and 2) NADPH reacts with methaemoglobin in the presence of NADPH-dependent methaemoglobin reductase and a cofactor to form haemoglobin. Low concentrations of methylene blue substitute for the cofactor, greatly increasing the activity of the pathway (Schimelman et al., 1978). Methylene blue reduces the half-life of methaemoglobin from 15–20 h to 40–90 min, acting as a cofactor to increase the erythrocytic production of methaemoglobin to oxyhaemoglobin in the presence of NADPH, generated by the hexose monophosphate shunt pathway; the methylene blue is oxidized to leukomethylene blue, which is the electron donor molecule for the non-enzymatic reduction of methaemoglobin to oxyhaemoglobin (Kumar et al., 1990).

In severe poisoning after ingestion, haemolytic anaemia may also occur on or about the fifth day. Its severity does not normally call for blood transfusion (Nabarro, 1948). In nitrobenzene poisoning, Heinz bodies are seen, as well as degenerating and regenerating forms of red blood cells, characterized by poikilocytosis and anisocytosis, many reticulocytes, polychromasia and many nucleated erythrocytes (Von Oettingen, 1941; David et al., 1965; Moeschlin, 1965).

Methaemoglobinaemia may be associated with sulfhaemoglobinaemia, arising as a result of interaction of sulfhydryl compounds with methaemoglobin (Von Oettingen, 1941); unlike methaemoglobin, sulfhaemoglobin formation is thought to be irreversible. In fatal poisoning cases, hyperplasia of the bone marrow has been reported (Zuccola, 1919).

In acute nitrobenzene poisoning, an increase in the number of leukocytes has been reported, with a relative lymphopenia (e.g., Von Oettingen, 1941; Parkes & Neill, 1953); in a protracted poisoning case, Carter (1936) reported leukopenia. No studies were located regarding immunological effects in humans after oral, dermal or inhalational exposure to nitrobenzene.

A general reduction in prothrombin activity, fibrinogen content, Factor VII and hyperfibrinolysis has been found in industrial workers after prolonged exposure (De Bruin, 1976).

Methaemoglobinaemia and haemolysis may lead to enlarging of the spleen (Ikeda & Kita, 1964).

8.4   Hepatic effects

Liver effects — i.e., low total plasma proteins with increased albumin/globulin ratio, a decrease in cholesterol esters and slight but significant bromosulfophthalein retention — were seen in a man who drank 15 ml of black laundry marking ink and 15 ml of denatured alcohol. These effects were ascribed to the nitrobenzene solvent in the ink (Wirtschafter & Wolpaw, 1944). The liver was enlarged and tender, and the results of liver function tests were abnormal (marked retention of bromosulfophthalein, slight increase in icterus index and indirect bilirubin) in a woman who was occupationally exposed to nitrobenzene vapour for 17 months (exposure levels not measured or estimated) (Ikeda & Kita, 1964). In the case of an attempted suicide reported by Parkes & Neill (1953) (see section 8.1.1) in which about 12 ml of nitrobenzene were ingested, excessive amino-aciduria indicated liver damage. Liver atrophy, with parenchymatous degenerative and necrotic foci, may be a sequela of severe poisoning (Von Oettingen, 1941).

8.5 Renal effects

Nitrobenzene poisoning may be associated with temporary anuria. When fatalities arise following extensive exposure, kidneys have been reported to show "cloudy swelling," parenchymatous degeneration and necrotic areas (Von Oettingen, 1941).

8.6 Neurological effects

Neurological effects following nitrobenzene ingestion by humans have been reported as headache, nausea, vertigo, confusion, unconsciousness, apnoea and coma (Leader, 1932; Carter, 1936; Myślak et al., 1971), as well as visual disturbances, reduced reflexes, spastic conditions, possibly even resulting in opisthotonus, tremors, twitching and convulsions (Von Oettingen, 1941). Neurological effects were also noted in a woman who was exposed to nitrobenzene vapours for 17 months at an unknown exposure level (Ikeda & Kita, 1964).

In severe poisoning, progressive drowsiness, coma and death from respiratory failure have been described (Chambers & O'Neill, 1945). The latter symptoms are ascribed to cerebral anoxia as well as a possible direct action of nitrobenzene on the nervous system. Oedema and hyperaemia of the meninges have been reported at necropsy of fatal cases (Von Oettingen, 1941). If cerebral anoxia is not promptly relieved, there is a risk of permanent damage of the basal ganglia (Adler, 1934; Locket, 1957). Permanent or long-term effects described include mental deterioration, rigour, catatonia and micrographia, restlessness and hyperkinesis, forgetfulness and memory disturbances (Grafe & Homburger, 1914; Von Oettingen, 1941; Moeschlin, 1965).

8.7 Other effects

No studies were located regarding possible carcinogenic, genotoxic or developmental effects of nitrobenzene in humans.

8.8 Subpopulations at special risk

Infants are especially vulnerable to methaemoglobinaemia because of the following factors (Von Oettingen, 1941; Goldstein et al., 1969):

- Fetal haemoglobin, which constitutes a proportion of the blood for some time after birth, is more prone to conversion to methaemoglobin than is adult haemoglobin.

- Because of a deficiency in the enzyme glucose-6-phosphate dehydrogenase in blood, infants are less able than older children and adults to reduce any methaemoglobin that might be formed in the blood back to haemoglobin.

In hereditary methaemoglobinaemia, the enzyme NADH methaemoglobin reductase is deficient, and persons are hypersensitive to any substances such as nitrite or aniline and aniline derivatives capable of producing methaemoglobinaemia (Goldstein et al., 1969). The trait is inherited as an autosomal recessive allele; thus, homozygotes may exhibit the trait, which is ordinarily detected by the presence of cyanosis at birth. Such individuals would be extremely sensitive to the effects of nitrobenzene, since they commonly exhibit levels of 10–50% methaemoglobinaemia. Although heterozygous individuals have no cyanosis, they are highly susceptible to oxidizing agents (Ellenhorn & Barceloux, 1988).

Genetically altered haemoglobins may confer abnormal sensitivity to compounds that cause methaemoglobinaemia, even though the normal reductive processes are operative. Thus, haemoglobin H, which consists of four β-chains instead of the usual two α- and two β-subunits, is unusually sensitive to oxidation, and methaemoglobin

accumulates in the older red blood cells, which become more susceptible to lysis. Haemoglobin M and haemoglobin S are susceptible
variants that differ from normal haemoglobin A in that one or more
amino acid residues have been replaced in the subunit chains
(Goldstein et al., 1969).

Several genetic variants of favism, leading to a decreased activity
of the enzyme glucose-6-phosphate dehydrogenase, are known
(Goldstein et al., 1969). This defect is ordinarily without adverse
effects, and it is only when these individuals are challenged with
compounds that oxidatively stress erythrocytes (e.g., primaquine) that
there is a haemolytic response. Reactors to primaquine (and fava
beans) have been reportedly found predominantly among groups that
live in or trace their ancestry to malaria-hyperendemic areas such as
the Mediterranean region or Africa. The incidence of "primaquine
sensitivity" among Kurds, a Middle Eastern population, is 53%.
Among blacks in the USA, the incidence is 13%. Individuals exhibiting such sensitivity would be expected to be more vulnerable to the
effects of nitrobenzene (Von Oettingen, 1941; Gosselin et al., 1984).

The mild analgesic and antipyretic agents acetanilide and phenacetin (now no longer used) can be metabolized to a minor extent by
deacetylation, yielding aniline derivatives that may form hydroxylaminobenzene and nitrosobenzene, compounds that can produce
methaemoglobin. Large and/or prolonged doses of phenacetin and
acetanilide have been shown to cause extensive methaemoglobin
formation (Carter, 1936; Goldstein et al., 1969). While paracetamol
(acetaminophen), a widely used antipyretic analgesic, is a major
metabolite of these two compounds, there is no evidence that methaemoglobinaemia is a side-effect of excessive paracetamol dosing
(Dukes, 1992), and thus its consumption is not likely to be a risk factor
for workers or others with the potential for significant exposure to
nitrobenzene.

Sulfonamide drugs, some of which are still used as anti-infective
drugs, have been reported to be able to cause methaemoglobinaemia
(Goldstein et al., 1969), as have the antimalarial drugs primaquine and
its 8-aminoquinoline congeners (Dukes, 1992); their use could possibly be a risk factor for anyone with the potential for significant
exposure to nitrobenzene.

Although the mechanism is not clear, a number of reports indicate that alcohol intensifies the methaemoglobinaemia (Gosselin et al., 1984) and other toxic symptoms (Von Oettingen, 1941) induced by nitrobenzene and aniline.

The employment of individuals with a predisposition to methaemoglobinaemia in industries manufacturing or using nitrobenzene and related compounds is of concern, since exposure to such compounds could put these individuals at risk (Linch, 1974).

9. EFFECTS ON OTHER ORGANISMS IN THE LABORATORY AND FIELD

9.1 Microorganisms

9.1.1 Toxicity to bacteria

The acute toxicity of nitrobenzene to the bacteria *Vibrio fischeri* (formerly *Photobacterium phosphoreum*) (Microtox) and *Pseudomonas putida* is given in Table 16. There was little variation in toxicity to *Vibrio fischeri* over 5- to 30-min exposures.

Blum & Speece (1991) studied the effect of nitrobenzene on bacterial populations. The inhibition of ammonia consumption was used as the criterion for *Nitrosomonas*, with an EC_{50} of 0.92 mg/litre. Inhibition of oxygen uptake was used as the criterion for aerobic heterotrophs, with an EC_{50} of 370 mg/litre. In anaerobic toxicity tests, the inhibition of gas production was used as the criterion for methanogens, with an EC_{50} of 13 mg/litre reported for nitrobenzene.

Yoshioka et al. (1986) studied the toxicity of nitrobenzene in the OECD activated sludge respiration inhibition test. The 3-h EC_{50} was 100 mg/litre. Using a modified OECD activated sludge respiration test with lower cell densities and substrate concentrations, Volskay & Grady (1988) found a 30-min EC_{50} of 320 mg/litre for nitrobenzene. Sealed vessels were used to overcome the problems of losses of volatile compounds.

9.1.2 Toxicity to protozoa

Bringmann & Kühn (1980) reported a toxic threshold (no-observed-effect concentration, or NOEC) for nitrobenzene of 1.9 mg/litre over a 72-h test period, based upon cell multiplication using the protozoan *Entosiphon sulcatum*.

Yoshioka et al. (1985) exposed the freshwater protozoan *Tetrahymena pyriformis* to nitrobenzene. A 24-h EC_{50}, based on growth

Table 16. Toxicity of nitrobenzene to bacteria

Bacteria	Effect parameter	Exposure duration (min)	Test end-point	Concentration (mg/litre)	Reference
Vibrio fischeri	Luminescence inhibition	15	EC$_{50}$	17.8	Deneer et al., 1989
	Luminescence inhibition	5	EC$_{50}$	28.2	Kaiser & Palabrica, 1991
	Luminescence inhibition	15	EC$_{50}$	29.5	Kaiser & Palabrica, 1991
	Luminescence inhibition	30	EC$_{50}$	34.7	Kaiser & Ribo, 1985
Pseudomonas putida	Cell multiplication inhibition	960	LOEC	7	Bringmann & Kühn, 1980

rate, was 98 mg/litre. Schultz et al. (1989) reported a 48-h EC_{50}, based on growth rate, of 106 mg/litre using the same species.

9.1.3   Toxicity to fungi

Gershon et al. (1971) studied the toxicity of nitrobenzene to a variety of fungi. Nitrobenzene showed fungistatic activity to *Myrothecium verrucaria* at 0.9 mg/litre and *Trichophyton mentagrophytes* at 0.5 mg/litre. For *Aspergillus niger, Aspergillus oryzae* and *Trichoderma viride*, there was no fungistatic effect at nitrobenzene concentrations of up to 1 mg/litre.

9.2   Aquatic organisms

Nitrobenzene is moderately toxic to microalgae, invertebrates, amphibians and fish. Most studies have investigated the toxicity of nitrobenzene to freshwater biota, with few data available for marine/ estuarine species. It should be noted that many studies carried out prior to 1985 did not incorporate adequate quality assurance procedures such as the use of criteria for test acceptability, reference toxicants and water quality monitoring (including the measurement of nitrobenzene) throughout the tests. Therefore, the data reported on the aquatic toxicity of nitrobenzene should be interpreted with caution.

9.2.1   Toxicity to algae

Toxicity of nitrobenzene to freshwater and marine microalgae is shown in Table 17. Based on inhibition of growth, reported 96-h EC_{50} values range from 17.8 to 43 mg/litre for freshwater algae. The only marine alga studied (*Skeletonema costatum*) was more sensitive to nitrobenzene, with a 96-h EC_{50} of 9.7 mg/litre.

9.2.2   Toxicity to invertebrates

Table 18 summarizes the acute and chronic toxicity of nitrobenzene to aquatic invertebrates. For freshwater invertebrates, 24- to 48-h LC_{50} values for nitrobenzene ranged from 24 mg/litre for the water flea (*Daphnia magna*) to 140 mg/litre for the snail (*Lymnaea stagnalis*). The flatworm (*Dugeis japonica*) was most sensitive, with a 168-h LC_{50} of 2 mg/litre. The only marine species tested was

Table 17. Toxicity of nitrobenzene to microalgae

Alga	Effect parameter	Exposure duration (h)	Test end-point	Concentration (mg/litre)	Reference
Blue-green					
Microcystis aeruginosa	Growth inhibition	192	LOEC	1.9	Bringmann & Kühn, 1978
Green					
Chlorella pyrenoidosa	Growth inhibition	72	EC$_{50}$	28 (23–36)[a]	Ramos et al., 1999
	Growth inhibition	72	LOEC	16	Ramos et al., 1999
	Growth inhibition	72	NOEC	9.2	Ramos et al., 1999
	Growth (cell yield) inhibition	96	EC$_{50}$	17.8	Deneer et al., 1989
Scenedesmus quadricauda	Growth inhibition	96	EC$_{50}$	40	Bringmann & Kühn, 1959
	Growth inhibition	192	LOEC	33	Bringmann and Kühn, 1980
Selenastrum capricornutum	Growth inhibition	96	EC$_{50}$	24 (9–39)	Bollman et al., 1990
	Growth (cell yield) inhibition	96	EC$_{50}$	43	US EPA, 1980
Diatom					
Skeletonema costatum[b]	Growth inhibition	96	EC$_{50}$	9.7	US EPA, 1980

[a] 95% confidence interval.
[b] Marine alga.

Table 18. Toxicity of nitrobenzene to aquatic invertebrates

Organism	Age (h)	Temper-ature (°C)	Hardness (mg CaCO₃/litre)	pH	Effect parameter	Exposure duration (days)	Test end-point	Concentra-tion (mg/litre)	Reference
Mysid shrimp[a] *Mysidopsis bahia*					Mortality	4	LC_{50}	6.7	US EPA, 1980
Water flea *Ceriodaphnia dubia*		24–26			Mortality	1	LC_{50}	54	Marchini et al., 1993
Water flea *Daphnia magna*		21–23		7.4–9.4	Mortality	1	LC_{50}	50	Bringmann & Kühn, 1982
	<24	21–23	173	7.4–9.4	Mortality	1	LC_{50}	24 (19–30)	LeBlanc, 1980
	<24	24–26			Immobilization	1	EC_{50}	60	Kühn et al., 1989
	<24	24–26			Immobilization	1	NOEC	19	Kühn et al., 1989
	<24	21–23	173		Mortality	2	LC_{50}	27 (22–32)	LeBlanc, 1980
					Mortality	2	LC_{50}	62	Canton et al., 1985
					Mortality	2	LC_{50}	33 (18–56)	Maas-Diepeveen & Van Leeuwen, 1986
					Immobilization	2	EC_{50}	35	Canton et al., 1985

157

Table 18 (Contd).

Organism	Age (h)	Temperature (°C)	Hardness (mg CaCO$_3$/litre)	pH	Effect parameter	Exposure duration (days)	Test endpoint	Concentration (mg/litre)	Reference
Water flea *Daphnia magna* (contd).		20	200	8.4	Immobilization	2	EC$_{50}$	33	Deneer et al., 1989
					Mortality	2	NOEC	0.46	LeBlanc, 1980
		24–26			Reproduction	14	NS[b]	12	Hattori et al., 1984
					Reproduction	21	NOEC	2.6	Kühn et al, 1989
Snail *Lymnaea stagnalis*					Mortality	1	LC$_{50}$	116 (98–139)	Ramos et al., 1998
					Mortality	2, 3 and 4	LC$_{50}$	64.5	Ramos et al., 1998
					Mortality	2	LC$_{50}$	140	Canton et al., 1985
Flatworm *Dugesis japonica*					Growth inhibition	7	EC$_{50}$	1.5	Yoshioka et al., 1986
					Mortality	7	LC$_{50}$	2	Yoshioka et al., 1986
Midge *Culex pipiens*					Mortality	2	LC$_{50}$	70	Canton et al., 1985

[a] Marine species.
[b] Not specified.

the mysid shrimp (*Mysidopsis bahia*), which was more sensitive than the freshwater species (96-h LC_{50} of 6.7 mg/litre).

In long-term toxicity tests (20 days) using *Daphnia magna*, Canton et al. (1985) reported 20-day values as follows: the LC_{50} was 34 mg/litre, the EC_{50} based on reproduction was 10 mg/litre, and the NOEC was 1.9 mg/litre. Similarly, Deneer et al. (1989) found a 21-day EC_{50} (based on immobilization in *Daphnia magna*) to be 24 mg/litre. The lowest concentration tested that significantly decreased the length of the daphnids was reported to be 17.8 mg/litre.

The reproductive toxicity of nitrobenzene in the water flea (*Daphnia magna*) was dependent on exposure duration, with NOEC values decreasing from 12 to 2.6 mg/litre over a 14- to 21-day exposure (Hattori et al., 1984; Kühn et al., 1989).

9.2.3 Toxicity to fish

Table 19 summarizes the acute toxicity of nitrobenzene to fish. The 96-h LC_{50} values for nitrobenzene ranged from 24 mg/litre for medaka (*Oryzias latipes*) to 142 mg/litre for guppy (*Poecilia reticulata*). One study (Yoshioka et al., 1986) reported that medaka were particularly sensitive, with a 48-h LC_{50} of 1.8 mg/litre.

Little information was available on long-term effects of nitrobenzene. Canton et al. (1985) reported an acute 18-day LC_{50} for nitrobenzene for medaka (*Oryzias latipes*) to be 24 mg/litre. The NOEC, based on mortality and behaviour, was 7.6 mg/litre.

Black et al. (1982) exposed rainbow trout (*Oncorhynchus mykiss*) embryo-larval stages (subchronic tests) from fertilization to 4 days post-hatching to nitrobenzene under flow-through conditions; total exposure time was 27 days. A wide range of concentrations was used in the test — 0.001, 0.01, 0.12, 0.36, 0.91 and 11.9 mg/litre. An LC_{50} of 0.002 mg/litre for nitrobenzene was reported at the time of hatching and at 4 days post-hatching. However, there is doubt about the validity of this figure, because concentrations below 0.12 mg/litre were nominal values.

Table 19. Acute toxicity of nitrobenzene to fish

Organism	Size/age	Temperature (°C)	Hardness (mg CaCO$_3$/litre)	pH	Effect parameter	Exposure duration (days)	Test end-point	Concentration (mg/litre)	Reference
Golden orfe *Leuciscus idus melanotus*					Mortality	2	LC$_{50}$	60–89	Juhnke & Lüdemann, 1978
Bluegill *Lepomis macrochirus*	0.32–1.2 g	21–23	32–48	6.7–7.7	Mortality	1	LC$_{50}$	135	Buccafusco et al., 1981
macrochirus	0.32–1.2 g	21–23	32–48	6.7–7.7	Mortality	4	LC$_{50}$	43 (36–49)[a]	Buccafusco et al., 1981
Fathead minnow *Pimephales promelas*	Larvae[b]	24–26			Mortality	4	LC$_{50}$	44	Marchini et al., 1992
	160 mg[b]	23–26	44.9	6.9–7.7	Mortality	1	LC$_{50}$	163	Holcombe et al., 1984
	160 mg[b]	23–26	44.9	6.9–7.7	Mortality	2	LC$_{50}$	156 (144–170)	Holcombe et al., 1984
	160 mg[b]	23–26	44.9	6.9–7.7	Mortality	3	LC$_{50}$	127	Holcombe et al., 1984

160

Table 19 (Contd).

Organism	Size/age	Temperature (°C)	Hardness (mg CaCO$_3$/litre)	pH	Effect parameter	Exposure duration (days)	Test endpoint	Concentration (mg/litre)	Reference
Fathead minnow Pimephales promelas (contd).	160 mg[b]	23–26	44.9	6.9–7.7	Mortality	4	LC$_{50}$	117	Holcombe et al., 1984
	30–35 days[b]	25			Mortality	4	LC$_{50}$	119	Schultz et al., 1989
			44	7.3	Mortality	4	LC$_{50}$	119 (107–133)	Geiger et al., 1985
	Larvae[b] (<24 h)	25	46		Mortality	4	LC$_{50}$	44 (41–48)	Marchini et al., 1992
	Larvae[b] (<24 h)	25	46		Mortality	7	LC$_{50}$	39 (36–42)	Marchini et al., 1992
	Larvae[b] (<24 h)	25	46		Mortality	7	LOEC	61	Marchini et al., 1992
	Larvae[b] (<24 h)	25	46		Mortality	7	NOEC	38	Marchini et al., 1992
	Larvae[b] (<24 h)	25	46		Growth	7	LOEC	<10	Marchini et al., 1992

Table 19 (Contd).

Organism	Size/age	Temper-ature (°C)	Hardness (mg CaCO₃/litre)	pH	Effect parameter	Exposure duration (days)	Test end-point	Concen-tration (mg/litre)	Reference
Sheepshead minnow	8–15 mm	25–31	10–31		Mortality	1, 2 and 3	LC_{50}	>120	Heitmuller et al., 1981
Cyprinodon variegatus	8–15 mm	25–31	10–31		Mortality	4	LC_{50}	59 (47–69)	Heitmuller et al., 1981
	8–15 mm	25–31	10–31		Mortality	4	NOEC	22	Heitmuller et al., 1981
Guppy *Poecilia reticulata*					Mortality	4	LC_{50}	142	Canton et al., 1985
					Mortality	4	LC_{50}	135 (121–150)	Ramos et al., 1998
					Mortality	14	LC_{50}	62	Maas-Diepeveen & Van Leeuwen, 1986; Deneer et al., 1987

Table 19 (Contd).

Organism	Size/age	Temperature (°C)	Hardness (mg CaCO$_3$/litre)	pH	Effect parameter	Exposure duration (days)	Test end-point	Concentration (mg/litre)	Reference
Medaka *Oryzias latipes*	0.2 g	25			Mortality	2	LC$_{50}$	20	Tonogai et al., 1982
	0.2 g	25			Mortality	4	LC$_{50}$	24	Canton et al., 1985
					Mortality	2	LC$_{50}$	1.8	Yoshioka et al., 1986
Zebra fish *Brachydanio rerio*				7.5	Mortality	4	LC$_{50}$	113	Wellens, 1982

[a] 95% confidence interval.
[b] Flow-through tests (all others were static tests).

9.2.4   *Toxicity to amphibians*

Canton et al. (1985) reported a 48-h LC_{50} for nitrobenzene in the South African clawed toad (*Xenopus laevis*) of 121 mg/litre and a 48-h EC_{50}, based on mortality and behaviour, of 54 mg/litre.

Black et al. (1982) exposed the leopard frog (*Rana pipiens*) to nitrobenzene from fertilization to 4 days post-hatching. Total exposure time was 9 days. A wide range of exposure concentrations was used — 0.001, 0.01, 0.05, 0.10, 0.41 and 1.27 mg/litre, with the concentrations below 0.10 mg/litre being nominal values. The LC_{50}, based on mortality at the time of hatching, was reported as >1.27 mg/litre, and at 4 days post-hatching, as 0.64 mg/litre.

9.3   Terrestrial organisms

9.3.1   *Toxicity to plants*

Fletcher et al. (1990) studied the effect of nitrobenzene on soybean (*Glycine max*) plants. The roots of the plant were exposed to a hydroponic solution containing ^{14}C-labelled nitrobenzene at 0.02–100 µg/litre for a 72-h exposure period. The chemical concentration in the solution was monitored, and the photosynthetic and transpiration rates were measured. The plants were dissected into roots and shoots and analysed for ^{14}C label and for nitrobenzene. There was no effect on transpiration or photosynthesis at the highest concentration tested (100 µg/litre), although root growth was inhibited.

9.3.2   *Toxicity to earthworms*

Neuhauser et al. (1985, 1986) studied the toxicity of organic compounds to the earthworm (*Eisenia foetida*), exposed via filter paper in glass vials in a 48-h contact test. A 48-h LC_{50} of 16 g/cm^2 filter paper was reported. Nitrobenzene was classified as very toxic to earthworms.

10. EVALUATION OF HUMAN HEALTH RISKS AND EFFECTS ON THE ENVIRONMENT

10.1 Evaluation of human health risks

10.1.1 Human exposure

10.1.1.1 General population exposure

General population exposure to nitrobenzene can result from releases to air and wastewater from industrial sources and from the presence of nitrobenzene as an air pollutant in ambient air, especially in urban areas in summer. Populations living in the vicinity of manufacturing activities involving nitrobenzene may receive significant exposure via ambient air. The occurrence of nitrobenzene in drinking-water is infrequent but possible following releases to water and soil. Occurrence in foods has not been reported. High concentrations are not likely, because the substance does not markedly bioaccumulate or biomagnify in the food-chain.

Because of nitrobenzene's ready detectability by both chemical analysis and human olfaction (sense of smell) and the relative ease of measurement of many of its properties, its release, transport, fate and the consequent exposure of human beings have been studied over a considerable period of time. Monitoring studies reveal low and highly variable exposures from air and, more rarely, drinking-water, with a generally downward trend in exposure levels. In some parts of the world, guideline values for nitrobenzene in ambient air have been set for residential areas.

Table 20 shows estimated daily absorbed doses of nitrobenzene derived from ambient air and drinking-water under normal backround (urban) conditions and in the worst-case situations arising from environmental contamination. Calculations are based on 24-h ventilaion of 22 m³ of air and 80% nitrobenzene retention in the lungs and on daily ingestion of 2 litres of water (IPCS, 1994; WHO, 1996), with complete absorption. It is realized that some additional nitrobenzene may be absorbed through the skin during washing, bathing and showering with nitrobenzene-contaminated water, but this route of exposure has not been quantified. Exposure of the general population to nitrobenzene via food or via consumer products cannot be estimated because pertinent data are lacking.

Table 20. Estimation of the general population exposure to nitrobenzene from ambient air and drinking-water

Source of exposure	Nitrobenzene concentration ($\mu g/m^3$ or $\mu g/litre$)	Daily absorbed dose ($\mu g/kg$ of body weight)
Air		
Normal urban air in summer	0.5	0.13
Urban area near industrial source	10	2.5
Drinking-water		
Normal background (upper end of the range)	0.7	0.02
Contaminated water (at the approximate level of odour threshold)	70	2

Under normal conditions, the combined exposure from air and drinking-water amounts to about 0.15 µg/kg of body weight per day. In the worst-case situation, populations living near an industrial emission source and using nitrobenzene-contaminated drinking-water with a concentration that can just be detectable by odour (by some people) might have a combined exposure amounting to a 30 times higher level (4.5 µg/kg of body weight per day).

10.1.1.2 Occupational exposure

Occupational exposure is of greatest concern, since nitrobenzene can be taken up very readily through the skin as well as by inhalation. Uptake of vapour through the skin (whole body) was estimated to be about one-fifth up to one-quarter of that occurring through the lungs via inhalation. Normal working clothes reduced skin absorption of vapour only by about 20–30%, based on the experiments by Piotrowski (1967).

Occupational exposure is likely to be significantly higher than general population exposure. Although a 1990 report from the USA estimated that over 10 000 workers were potentially exposed to nitrobenzene, there are few quantitative data on exposure levels.

In old studies of a factory producing nitrobenzene, average air concentrations were reported as 196 mg/m^3 in 1952 and 29 mg/m^3 in 1954. Both values were associated with symptoms in workers. A more recent study by Harmer et al. (1989) found airborne levels of nitrobenzene generally <5 mg/m^3 and urinary *p*-nitrophenol levels generally below 5 mg/litre. In workplaces with good controls, exposures should be below the occupational exposure limits of 5 mg/m^3 adopted by 28 countries (listed in IARC, 1996). All occupational exposure limits have an additional "skin" notation to warn of the potential for significant absorption of nitrobenzene across the skin and of the need to prevent it. As an additional aid to exposure assessment, some national jurisdictions (e.g., the USA and Germany) have proposed biological monitoring guidance values to be used as biomarkers.

In the USA, two methods for biological monitoring of nitrobenzene exposures have been used: total *p*-nitrophenol in urine (Lauwerys, 1991; ACGIH, 1995) and methaemoglobin in the blood (ACGIH, 1995). The latter biomarker is less specific for nitrobenzene exposure, since a number of other redox chemicals can cause methaemoglobinaemia. Analysis of aniline–haemoglobin conjugate is recommended for the biological monitoring of nitrobenzene in Germany (Neumann, 1988; DFG, 1995).

10.1.2   Hazard identification

At moderate to high doses in experimental animals, nitrobenzene can cause methaemoglobinaemia, haemolytic anaemia, medullary and extramedullary haematopoiesis and toxic effects in the spleen, liver and kidney. Further effects include testicular toxicity and impairment of male fertility, neurological effects and toxic effects on the thyroid, adrenal gland and immune system. Long-term studies have indicated local effects on the upper airways and lungs.

10.1.2.1   Death

Numerous accidental poisonings and deaths in humans that were attributed to the ingestion of nitrobenzene have been reported. In cases of oral ingestion or inhalation in which the patients were apparently near death due to severe methaemoglobinaemia, termination of exposure and prompt medical intervention resulted in gradual improvement and recovery. Some early data relating to dermal exposures may relate to aniline rather than nitrobenzene, but there are reports of deaths arising from nitrobenzene spilled on clothing.

10.1.2.2 Methaemoglobinaemia

The most commonly reported systemic effect associated with human exposure to nitrobenzene is methaemoglobinaemia, a haemopathy exhibiting an oxidized form of haemoglobin. As a normal blood constituent, methaemoglobin is usually kept low by the reduction of haem Fe^{3+} to Fe^{2+} by methaemoglobin reductase in healthy, unexposed individuals.

Oral exposure to nitrobenzene in unspecified amounts resulted in methaemoglobinaemia. Clinical reports of methaemoglobinaemia following exposure to nitrobenzene via inhalation include twin 3-week-old babies, a 12-month-old girl and a 47-year-old woman. However, levels of exposure were neither known nor estimated. Methaemoglobinaemia has been reported after dermal exposure in humans; in fact, the occurrence of cyanotic babies in hospitals that used diapers stamped with inks using nitrobenzene or aniline solvents appears to have been relatively common in earlier times. Methaemoglobinaemia has also been seen in animals exposed to nitrobenzene via the oral, inhalation and dermal routes.

10.1.2.3 Splenic effects

The spleen is a target organ during human and animal exposure to nitrobenzene. For example, a 47-year-old woman occupationally exposed to nitrobenzene in paint (mainly by inhalation) had a palpable and tender spleen.

Splenic lesions have also been reported following nitrobenzene inhalation in mice and rats. Splenic lesions seen were sinusoidal congestion, an increase in extramedullary haematopoiesis and haemosiderin-laden macrophages invading the red pulp, and the presence of proliferative capsular lesions.

10.1.2.4   *Hepatic effects*

Liver effects have been reported in humans exposed to nitro-benzene. Hepatic enlargement and tenderness and altered serum chemistries were reported in a woman who had been occupationally exposed to nitrobenzene. The authors considered these changes to be related to increased destruction of haemoglobin and red blood cells and enlargement of the spleen. However, an extensive range of liver pathologies in animal studies suggests a possibility of more direct target organ toxicity.

10.1.2.5   *Renal effects*

There are no data on renal effects in humans exposed to nitro-benzene by any route. Observed effects in laboratory animals have included increased kidney weights, pigmentation of tubular epithelial cells, hydropic degeneration of the cortical tubules and protein nephropathy, and swelling of the glomeruli and tubular epithelium. The available evidence is not sufficient to indicate whether the renal effects are specific to male rats. The implications for humans are unclear.

10.1.2.6   *Immunological effects*

No studies were located regarding immune system effects in humans after exposure to nitrobenzene.

In an immunotoxicity study in mice, immunosuppression was evidenced by a diminished IgM response to sheep red blood cells. Host resistance to microbial or viral infection was not markedly affected by nitrobenzene, although there was a trend towards increased susceptibility in cases in which T-cell function contributes to host defence. In a long-term study with mice, thymic involution was found.

10.1.2.7   Neurological effects

Neurotoxic symptoms reported in humans after inhalation exposure to nitrobenzene have included headache, confusion, vertigo and nausea. Effects in orally exposed persons have also included those symptoms, as well as apnoea and coma.

Damage to the brain stem, cerebellum and fourth ventricle was observed in orally exposed animals, whereas animals exposed via inhalation have shown morphological damage to the hind brain.

10.1.2.8   Reproductive effects

Effects of nitrobenzene on reproduction in humans have not been reported.

In Sprague-Dawley rats, inhalation of nitrobenzene in a two-generation study resulted in decreased male fertility, with alterations in male reproductive organs (e.g., seminiferous tubular atrophy and spermatocyte degeneration) with at least partial functional reversibility after a 9-week recovery period. In Fischer-344 rats, no recovery of cessation of spermatogenesis after a 14-day recovery period could be observed in a short-term inhalation study. Furthermore, oral and dermal application of nitrobenzene to different rat and mouse strains also led to impaired male fertility, with alterations in male reproductive organs. Nitrobenzene has direct effects on the testis, shown by *in vivo* and *in vitro* studies. Spermatogenesis is affected, with exfoliation of predominantly viable germ cells and degenerating Sertoli cells. The main histopathological effects are degenerated spermatocytes.

In general, maternal reproductive organs were not affected; only one study showed uterine atrophy in mice after dermal application of high doses.

No studies of developmental effects in humans resulting from exposure to nitrobenzene have been reported.

Studies conducted via oral or inhalation exposure did not result in fetotoxic or teratogenic effects in rats or rabbits. No studies have been conducted using the dermal route.

10.1.2.9 Genotoxic effects

The genotoxicity of nitrobenzene has been evaluated in both *in vitro* and *in vivo* studies. The results of these studies are generally negative.

Nitrobenzene was extensively studied for the induction of gene mutations in bacteria. The results of these studies were principally negative. Only in one case, in the presence of a co-mutagen, was a positive response observed. The ability of nitrobenzene to induce gene mutations in mammalian cells was tested in only one study. A weak positive result was observed (mutation to 8-azaguanine-resistance), or the positive effect was only marginal (mutation to ouabain-resistance). One study showed a weak increase in recessive lethal mutations in *Drosophila*; however, lack of study details made it difficult to interpret the result. Nitrobenzene did not induce DNA repair or chromosome damage in mammalian cells either *in vitro* or *in vivo*.

Nitrobenzene did not induce cell transformation in cultured Syrian hamster kidney cells or human lung fibroblasts.

One *in vitro* study suggested binding to DNA. An *in vivo* study in rats and mice given subcutaneous ^{14}C-labelled nitrobenzene demontrated covalent binding of radioactivity to DNA isolated from rat liver and kidney and mouse liver and lung. The binding was within the range typically found with weak genotoxic carcinogens. Furthermore, there is only limited evidence for the genotoxicity of nitrobenzene metabolites. Available genotoxicity data do not suggest potential gentic effects of concern in humans.

10.1.2.10 Carcinogenic effects

A 2-year inhalation study showed that $B6C3F_1$ mice, Fischer-344 rats and Sprague-Dawley CD rats can respond with tumours at eight different organ sites. Three of the eight sites responded with significant evidence of carcinogenicity: 1) mammary adenocarcinomas in female $B6C3F_1$ mice; 2) liver carcinomas in male Fischer-344 rats; and 3) thyroid follicular cell adenocarcinomas in male Fischer-344 rats. The rest of the sites (five organs) were mostly benign tumorigenic responses (Table 21).

Nitrobenzene can be reduced by gut microflora and, after system absorption, by cellular microsomes, forming the carcinogenic metabolites nitrosobenzene, phenylhydroxylamine and aniline. Nitrobenzene can also be systemically oxidized by the cellular cytochrome-450 microsomal system to various chemically reactive nitrophenols. Although all these redox metabolic products are candidates for cancer causality, the mechanism of carcinogenic action is not known. Because of the likely commonality of redox mechanisms producing similar chemical intermediates in test animals and humans, it is hypothesized that nitrobenzene may cause cancer in humans by any route of exposure.

Table 21. Summary of nitrobenzene carcinogenicity results from section 7.3[a]

Site of increased tumorigenicity	Sex	Evidence of carcinogenicity	Comments
$B6C3F_1$ mouse			
Lung: Alveolus and bronchus	m	Limited	Benign tumour increase only; carcinomas spread evenly over dose groups (no trend)
Thyroid: Follicular cell	m	Limited	Benign tumours only with dose trend
Mammary gland	f	Sufficient (lacks low- and mid-dose results)	Historical and concurrent controls suggest malignant but low-level cancer incidence in comparison of 0 and 260 mg/m^3 (0 and 50 ppm) dose groups

Table 21 (Contd).

Fischer-344 rat

Liver: Hepatocellular	m	Sufficient	Clear evidence of malignancy and exposure related to cancer incidence in the liver
Thyroid: Follicular cell	m	Limited	Another thyroid follicular response; marginal statistics: only a trend with dose and only a suggestion of malignancy
Kidney: Tubular cell	m	Limited	High dose only; benign tumorigenic response
Endometrial polyp	f	Limited	Benign response
CD rat			
Liver: Hepatocellular	m	Limited	Benign response (unlike the male Fischer-344 rat)

[a] From Holder (1998, 1999a).

10.1.3 *Dose–response analysis*

10.1.3.1 *Non-neoplastic effects*

Available human data are too limited to allow detailed comments to be made about the relationship between the level of exposure and the degree of toxic response. It is clear from the literature that recovery from short-term exposures to low concentrations can occur without medical intervention, but that exposures to higher concentrations by any route require medical intervention. It would appear from an overview of the human poisoning literature that severity of effect is indeed related to the degree of systemic exposure. One problem with a number of reports in the medical literature is that nitrobenzene was ingested in products containing a "cocktail" of other ingredients or that its intake could have been confused with intake of aniline.

For the oral, dermal and inhalation routes, adequate toxicology studies are available for mice and rats. They all reveal similar systemic effects. In none of the studies did the findings allow a NOAEL to be established. Methaemoglobinaemia, haematological and testicular effects and, in the inhalation studies, effects on the respiratory system were found at the lowest doses tested (Tables 22–24).

Table 22. LOAELs[a] for non-neoplastic effects in rats and mice in oral, dermal and inhalation studies with nitrobenzene

Study details			LOAEL[a]	
Species/strain/sex	Route	Duration	ppm	mg/kg of body weight per day
Rats — F344 (m and f)	Oral	13 weeks		9.375
	Dermal	13 weeks		50
	Inhalation	2 years	1	0.7[b]
Rats — CD (m)	Inhalation	2 years	1	0.7[b]
Mice — B6C3F$_1$ (m and f)	Oral	13 weeks		18.75
	Dermal	13 weeks		50
	Inhalation	2 years	5	4.3[c]

[a] The lowest doses tested in these GLP studies were taken to be the LOAELs. Note that 1 ppm = 5.12 mg/m^3.
[b] Systemic dose from inhalation exposure calculated assuming a respiratory volume of 0.223 m^3/day, a body weight of 0.3 kg and 70% retention in the lung.
[c] Systemic dose from inhalation exposure calculated assuming a respiratory volume of 0.039 m^3/day, a body weight of 0.04 kg and 70% retention in the lung.

In chronic inhalation studies, **methaemoglobinaemia** was consistently observed at 130 mg/m^3 (25 ppm) and above in B6C3F$_1$ mice and rats (both Sprague-Dawley and Fischer-344 strains) at the 2-year terminal sacrifice, but it was also apparent at the lowest dose of 5 mg/m^3 (1 ppm) at interim (15-month) sacrifice in both strains of rats, with Sprague-Dawley rats more affected at this dose than the Fischer-344 strain.

Table 23. Incidence of chronic non-neoplastic lesions in B6C3F1 mice following nitrobenzene exposure[a]

Tissue/diagnosis	Sex	Incidence (%) at concentration (ppm)[b]			
		0	5	25	50
Lung					
Alveolar/bronchial hyperplasia	m	1/68 (1)*T	2/67 (3)	8/65 (12)*	13/66 (20)*
	f	0/53 (0)	2/60 (3)	5/64 (8)*	1/62 (2)
Bronchiolization of alveolar walls	m	0/68 (0)*T	58/67 (87)*	58/65 (89)*	62/66 (94)*
	f	0/53 (0)*T	55/60 (92)*	63/64 (98)*	62/62 (100)*
Thyroid gland					
Follicular cell hyperplasia	m	1/65 (2)*T	4/65 (6)	7/65 (11)*	12/64 (19)*
	f	2/49 (4)*T	1/59 (2)	1/61 (2)	8/61 (13)
Liver					
Centrilobular hepatocytomegaly	m	1/68 (1)*T	15/65 (23)˙	44/65 (68)*	57/64 (89)*
	f	0/51 (0)*T	0/61 (0)	0/64 (0)	7/62 (11)*
Multinucleated hepatocytes	m	2/68 (3)*T	14/65 (22)˙	45/65 (69)*	50/64 (88)*
	f	0/51 (0)	0/61 (0)	0/64 (0)	2/62 (3)

Table 23 (Contd).

Tissue/diagnosis	Sex	Incidence (%) at concentration (ppm)[b]				
		0	5	25	50	
Nose						
Glandularization of respiratory epithelium	m	10/67 (15)*[T]	0/66 (0)	0/65 (0)	27/66 (41)*	
	f	0/52 (0)*[T]	0/60 (0)	0/63 (0)	7/61 (11)*	
Increased secretion from respiratory epithelium	m	0/67 (10)*[T]	0/66 (0)	3/65 (5)	6/66 (9)*	
	f	2/52 (4)*[T]	7/60 (12)	19/63 (30)*	32/61 (52)*	
Degeneration/loss of olfactory epithelium	m	1/67 (1)*[T]	1/66 (2)	32/65 (49)*	41/66 (62)*	
	f	0/52 (0)*[T]	19/60 (32)*	47/63 (75)*	42/61 (69)*	
Pigment in olfactory epithelium	m	0/67 (0)*[T]	7/66 (11)*	46/65 (71)*	49/66 (74)*	
	f	0/52 (0)*[T]	6/60 (10)*	37/63 (59)*	29/61 (48)*	
Testes						
Diffuse atrophy	m	1/68 (1)	2/5 (40)	0/7(0)	6/66 (9)	
Decreased spermatogenesis	m	1/68 (1)	0/5 (0)	0/7(0)	3/66 (5)	
Tubular atrophy	m	9/68 (13)	0/5 (0)	1/7 (14)	1/66 (2)	

176

Table 23 (Contd).

Tissue/diagnosis	Sex	Incidence (%) at concentration (ppm)[b]			
		0	5	25	50
Epididymis					
Hypospermia	m	3/68 (4)	2/5 (40)	0/6 (0)	11/66 (17)
Bone marrow, femur					
Hypercellularity	m	3/68 (4)[*T]	10/67 (15)*	4/64 (6)	13/66 (20)*
	f	4/52 (8)	–	–	9/62 (15)
Thymus					
Involution	m	10/48 (21)	–	–	10/44 (23)
	f	7/41 (17)	–	–	22/57 (39)*
Kidney					
Cysts	m	2/68 (3)	–	–	12/65 (18)*
	f	0/51 (0)	–	–	0/62 (0)
Pancreas					
Mononuclear cell infiltration	m	3/65 (5)	–	–	3/64 (5)
	f	1/46 (2)	–	–	8/62 (13)*

[a] From Cattley et al. (1994).
[b] * = significant at $P < 0.05$; *T = significantly positive correlation trend, $P < 0.5$; 1 ppm = 5.12 mg/m^3.

Table 24. Incidence of chronic non-neoplastic lesions in Fischer-344 and CD rats following nitrobenzene exposure[a]

Tissue/diagnosis	Species	Sex	Incidence (%) at concentration (ppm)[b]			
			0	1	5	25
Liver						
Eosinophilic foci	CD	m	11/63 (17)[*T]	3/67 (4)	8/70 (11)	19/65 (29)
	F344	m	26/69 (42)[*T]	25/69 (36)	44/70 (63)[*]	57/70 (81)[*]
		f	6/70 (9)[*T]	9/66 (14)	13/66 (20)	16/70 (23)[*]
Centrilobular hepatocytomegaly	CD	m	3/63 (5)[*T]	1/67 (1)	14/70 (20)[*]	39/65 (60)[*]
	F344	m	0/69(0)[*T]	0/69 (0)	8/70 (11)[*]	57/70 (81)[*]
		f	0/70 (0)	0/66 (0)	0/66 (0)	0/70 (0)
Spongiosis	CD	m	25/63 (40)[*T]	25/67 (37)	25/70 (36)[*]	37/65 (57)[*]
	F344	m	25/69(36)[*T]	24/69 (35)[T]	33/70 (47)	58/70 (83)[*]
		f	0/70 (0)[*T]	0/66 (0)	0/66 (0)	6/70 (9)[*]
Kidney						
Chronic nephropathy	CD	m	54/63 (86)	60/67 (90)	63/70 (90)	59/65 (91)
	F344	m	69/69 (100)	64/68 (94)	70/70 (100)	70/70 (100)

Table 24 (Contd).

Tissue/diagnosis	Species	Sex	Incidence (%) at concentration (ppm)[b]			
			0	1	5	25
Chronic nephropathy (contd).	F344	f	58/70 (83)	51/66 (77)	60/66 (91)	67/70 (96)
Tubular hyperplasia	CD	m	3/63 (5)	1/67 (1)	5/70 (7)	6/65 (9)
	F344	m	2/69 (3)*T	2/68 (3)	2/70 (3)	13/70 (19)*
		f	0/70 (0)	0/66 (0)	2/66 (3)	2/70 (3)
Thyroid gland						
Follicular cell hyperplasia	CD	m	2/63 (3)	2/64 (3)	1/68 (1)	4/64 (6)
	F344	m	0/69 (0)*T	1/69 (1)	2/70 (3)	4/70 (6)
		f	1/69 (1)	–	–	0/68 (0)
Nose						
Pigment deposition in olfactory epithelium	CD	m	42/63 (67)*T	49/64 (77)	60/66 (91)*	58/61 (95)*
	F344	m	40/67 (60)*T	53/67 (79)*	67/70 (96)*	68/69 (99)*
		f	37/67 (55)*T	54/65 (83)*	60/65 (92)*	66/66 (100)*
Testes						
Bilateral atrophy	CD	m	11/62 (18)*T	17/66 (26)	22/70 (31)	35/61 (57)*

Table 24 (Contd).

Tissue/diagnosis	Species	Sex	Incidence (%) at concentration (ppm)[b]			
			0	1	5	25
Bilateral atrophy (contd).	F344[c]	m	61/69 (88)	50/56 (89)	59/61 (97)	61/70 (87)
Epididymis						
Bilateral hypospermia	CD	m	8/60 (13)[*T]	13/65 (20)	15/67 (22)	32/59 (54)[*]
	F344	m	15/69 (22)	21/54 (39)	12/59 (20)	12/70 (17)
Spleen						
Extramedullary haematopoiesis	CD	m	58/63 (92)	56/67 (84)	61/69 (88)	60/65 (92)
	F344	m	53/69 (77)	62/69 (90)[*]	65/70 (93)[*]	61/70 (87)
		f	60/69 (87)	62/66 (94)	60/66 (91)	65/69 (94)
Pigmentation	CD	m	59/63 (94)[*T]	58/67 (87)	67/69 (97)	65/65 (100)
	F344	m	55/69 (80)[*T]	63/69 (91)[*]	64/70 (91)[*]	70/70 (100)[*]
		f	62/69 (90)[*T]	61/66 (92)	60/66 (91)	68/69 (99)[*]

[a] From Cattley et al. (1994)

[b] * = significant at $P < 0.05$; *T significantly positive correlation trend, $P < 0.05$; 1 ppm = 5.12 mg/m^3.

[c] Data for F344 rats are listed as unilateral or bilateral.

In mice, **testicular effects** were adequately investigated only at the high dose (260 mg/m³ [50 ppm]), with bilateral epididymal hypospermia noted with an incidence of 17% (compared with 4% in controls). In Sprague-Dawley but not Fischer-344 rats, bilateral epididymal hypospermia was increased from the lowest dose of 5 mg/m³ (1 ppm) (significant positive trend, $P < 0.05$). In mice, diffuse testicular atrophy was seen at the highest dose (260 mg/m³) (the only group examined). In Sprague-Dawley but not Fischer-344 rats, there was a dose-related increase in bilateral atrophy, apparent from the lowest dose (5 mg/m³) (significant positive trend, $P < 0.05$).

In mice, there was a dose-related increase in the incidence of **alveolar/bronchial hyperplasia**, with a small increase compared with controls even at the lowest dose tested of 26 mg/m³ (5 ppm) (apparent in both sexes, positive trend in males at $P < 0.5$). Nitrobenzene also caused bronchiolization of alveolar walls in almost all exposed mice — i.e., it was apparent from the lowest exposure level of 26 mg/m³. Similar findings were not reported in rats of either strain. A range of pathologies was also seen in the olfactory epithelium of mice from the lowest dose of 26 mg/m³, including "increased secretory product" (females), degeneration (females) and pigmentation (both sexes). In both strains of rats, dose-related pigmentation of the olfactory epithelium was reported, apparent from the lowest dose of 5 mg/m³ (1 ppm) (significant positive trend, $P < 0.05$).

Other findings, apparent at the lowest inhalational dose, included extramedullary haematopoiesis and pigment deposition in the spleen (male Fischer-344 rats), liver pathology (mice and female Fischer-344 rats) and bone marrow hypercellularity (mice).

10.1.3.2   *Carcinogenic effects*

Dose–response considerations with respect to tumorigenic and carcinogenic effects are presented in Appendix 1. Note that these quantitative risk estimate calculations have been performed using methods generally employed by the US EPA and are not methods adopted or used by IPCS. Estimated inhalation exposure, for which the upper 95% confidence limit for the cumulative lifetime risk is 10^{-6}, is approximately 10 ng/kg of body weight daily.

10.1.4   Risk characterization

10.1.4.1   General population

Exposure of the general population to nitrobenzene from air or drinking-water is likely to be very low. Although no NOAEL could be derived from any of the toxicological studies, there is a seemingly low risk for non-neoplastic effects. If exposure values are low enough to avoid non-neoplastic effects, it is expected that carcinogenic effects will not occur.

No information is available to indicate whether nitrobenzene is still used in consumer products; thus, the extent of exposure to nitrobenzene from these sources could not be quantified. As described in many published studies, poisoning from the use of such consumer products has occurred frequently in the past. Significant human exposure is possible, due to the moderate vapour pressure of nitrobenzene and extensive skin absorption. Furthermore, the relatively pleasant almond smell of nitrobenzene may not discourage people from consuming food or water contaminated with it. Infants are especially susceptible to the effects of nitrobenzene.

10.1.4.2   Workplace

There is limited information on exposure in the workplace. In one workplace study, no effect on methaemoglobin levels was noted at exposure concentrations ranging from 0.7 to 2.2 mg/m³. However, in this study, exposure concentrations were of the same order of magnitude as the LOAELs in a long-term inhalation study, a study that also demonstrated that nitrobenzene is a carcinogen in different species and strains in several organs. Therefore, there is significant concern for the health of workers exposed to nitrobenzene, due to its toxic and carcinogenic effects.

10.2 Evaluation of effects on the environment

10.2.1   Exposure

Nitrobenzene's moderate volatility and weak sorption to soil suggest that it may have the potential to contaminate surface water and groundwater. However, environmental levels are mitigated to some extent by degradation, including photolysis and microbial biodegradation. Concentrations of nitrobenzene in environmental samples such as surface waters, groundwaters and air are generally low.

Measured concentrations for nitrobenzene in urban air samples in summer range from <0.05 to 2.1 $\mu g/m^3$ (<0.01 to 0.41 ppb), with slightly higher concentrations found in industrial areas. Studies of municipal waste disposal facilities and hazardous waste sites have found nitrobenzene infrequently present in air releases and, if detected, generally at low concentrations. Air levels were significantly lower (or undetectable) in winter than in summer, due to both the formation of nitrobenzene by nitration of benzene (from petrol) and the higher volatility of nitrobenzene during the warmer months.

Typical concentrations of nitrobenzene in surface waters range from 0.1 to 1 $\mu g/litre$; however, concentrations of up to 67 $\mu g/litre$ were reported in the river Danube, Yugoslavia, in 1990. Groundwater concentrations ranging from not detected (detection limit 1.13 $\mu g/litre$) to 4.2 mg/litre have been reported from three sites in the USA. Based on limited data, it appears that there may be potential for contamination of groundwaters, with the highest concentrations being reported at hazardous waste sites and near coal gasification sites.

Nitrobenzene in drinking-water has been reported in studies conducted in the 1970s and 1980s in the USA and the United Kingdom, albeit in only a small proportion of samples, but was not detected in 30 Canadian samples (1982 report). There has generally been a downward trend in the concentrations of nitrobenzene found in drinking-water over the past two decades.

Nitrobenzene shows a low tendency for adsorption onto soils and sediments and is likely to be highly mobile in such media. It is intermediately mobile in forest and agricultural soils, being somewhat more mobile in soil with lower organic content. Nitrobenzene may also be present in soils at hazardous waste sites.

The measured BCFs for nitrobenzene in a number of organisms indicate minimal potential for bioaccumulation, and nitrobenzene is not biomagnified through the food-chain. It was not detected in a large range of sampled biota in a 1985 US study. Nitrobenzene can be taken up by plants. In available studies, it appeared to be associated with roots, and very little was associated with other parts of the plant.

Nitrobenzene can undergo degradation by both photolysis and microbial biodegradation. A number of fairly stable degradation products of nitrobenzene are formed during environmental degradation; some have toxic effects that are similar to those of nitrobenzene, whereas others have different modes of action. Whether or not nitrobenzene will be completely broken down (mineralized) at a particular site seems to be questionable.

Nitrobenzene is slowly photolysed in air and water. The estimated lifetime for the direct photolysis of airborne nitrobenzene was <1 day. UV photolysis may be a pathway for nitrobenzene degradation in natural waters, but only where biodegradation is slow and where water is relatively clear. Reaction of nitrobenzene with hydroxyl radicals is likely to be only a minor removal pathway for nitrobenzene in natural waters. Half-lives in water for the direct photolysis reaction were calculated as 2.5–6 days. Indirect photolysis (photo-oxidation) plays a minor role in both water and air. Nitrobenzene does not deplete the stratospheric ozone layer, and no information has been reported on its global warming potential.

Degradation studies suggest that nitrobenzene is degraded in sewage treatment plants by aerobic processes, with slower degradation under anaerobic conditions. Nitrobenzene may not necessarily be completely degraded if present at high concentrations in wastewater. High concentrations may also inhibit the biodegradation of other wastes. Biodegradation of nitrobenzene depends mainly on the acclimation of the microbial population. Degradation by non-acclimated inocula is generally very slow to negligible and proceeds only after extended acclimation periods. Acclimated microorganisms, particularly from industrial wastewater treatment plants, however, showed complete elimination within a few days. Degradation was generally found to be increased in the presence of other easily degradable substrates. Adaptation of the microflora and additional substrates also seem to be limiting factors for the decomposition of nitrobenzene in soil. Degradation of nitrobenzenes under anaerobic conditions has been shown to be very slow, even after extended acclimation periods.

10.2.2   Effects

Nitrobenzene is of moderate to low toxicity to aquatic and terrestrial organisms. Reported LC_{50}/EC_{50} data for acute toxicity to freshwater organisms range from 2 to 156 mg/litre. The lowest acute NOEC was 0.46 mg/litre for the water flea (*Daphnia magna*); however, this value should be interpreted with caution, as it was lower than the chronic NOEC for the same species. No effects of nitrobenzene have been reported at concentrations lower than 1 mg/litre for other invertebrates and fish. Data on marine organisms are limited to one fish (sheepshead minnow, *Cyprinodon variegatus*), which showed similar sensitivity to freshwater species, one alga (*Skeletonema costatum*) and one crustacean (mysid shrimp, *Mysidopsis bahia*), which were more sensitive to nitrobenzene than freshwater species.

Little information was available on the chronic toxicity of nitrobenzene to freshwater biota. A NOEC of 1.9 mg/litre was reported for chronic studies (20-day exposure) with the water flea (*Daphnia magna*).

Nitrobenzene appears to be toxic to bacteria and may adversely affect sewage treatment facilities if present in high concentrations in influent. For terrestrial systems, the reported levels of concern in laboratory toxicity tests are unlikely to occur in the natural environment, except possibly in areas close to nitrobenzene production and use and in areas contaminated by spillage.

10.2.3   Risk characterization

Because of the low toxicity and low bioaccumulation potential of nitrobenzene, it is unlikely to pose an environmental hazard at typical concentrations found in environmental samples. There is little overlap between typical concentrations in environmental samples and those concentrations known to be toxic to aquatic and terrestrial organisms.

For nitrobenzene, the most sensitive species was reported to be the embryo-larval stages of rainbow trout (*Oncorhynchus mykiss*), with an LC_{50} at 0.002 mg/litre after a 27-day exposure. However, because there is some doubt as to the validity of this result, this datum was not used in this risk assessment.

An environmental concentration of concern may be calculated using the next most sensitive species — i.e., the mysid shrimp (*Mysidopsis bahia*) — with a 96-h LC_{50} of 6.7 mg/litre. However, this is an acute value, and it is preferable to use chronic test data. The lowest chronic test datum available is a 20-day NOEC for *Daphnia magna* of 1.9 mg/litre. Application of an assessment factor of 100 gives an environmental concentration of concern of 20 µg/litre. However, the use of such conservative assessment (safety) factors for the derivation of guidance values has recently been questioned (Chapman et al., 1998).

A preferable approach is to fit the species sensitivity distributions to a logistic (or more general Burr Type III) distribution to derive a hazardous concentration to protect 95% of species with an associated confidence level (HC_5) (Aldenberg & Slob, 1993). Using the available acute toxicity data for nitrobenzene and this statistical distribution method, together with an acute:chronic ratio of 16 derived from the crustacean data, a moderate reliability freshwater guidance value was derived. To protect 95% of freshwater species with 50% confidence, the concentration limit for nitrobenzene was estimated to be 200 µg/litre. The description of this method (statistical extrapolation method) is presented in Appendix 2.

There were insufficient data to derive a marine guideline value.

11. RECOMMENDATIONS FOR PROTECTION OF HUMAN HEALTH

11.1 Public health

The following recommendations are made for the protection of human health:

- Avoid consumer exposure to nitrobenzene by its removal from consumer products wherever possible.

- Any product containing nitrobenzene should be clearly labelled.

- Nitrobenzene must not be used for products that are formulated for application to the skin or that may come into contact with the skin.

- Steps that guarantee minimal emissions of nitrobenzene at production sites should be implemented and enforced worldwide.

- Contaminated areas such as landfill sites and injection wells for spent nitrobenzene and related chemicals should be identified, and appropriate surveillance and introduction of measures to control emissions to air and water should be instituted.

Population groups particularly at risk and needing additional protection include:

- infants and children;
- pregnant women;
- lactating women;
- people with enzyme deficiencies leading to a predisposition to methaemoglobinaemia; and
- people with haemoglobin variants (H, M and S).

11.2 Occupational health

Since nitrobenzene is a potent toxicant and a potential carcinogen, exposures should be kept as low as possible, using the best available technology worldwide.

Nitrobenezene liquid can be absorbed through the skin, and nitrobenzene vapour can be absorbed through the lungs and the skin. Particular attention must be given to prevention of skin contact by both nitrobenzene liquid and vapour by adequate protective equipment. In order to minimize exposure, the application of a valid method of biological monitoring is recommended.

Workers potentially exposed to nitrobenzene require adequate safety information, education and training regarding the risks of nitrobenzene exposure and the need to adopt safe working procedures.

Monitoring of nitrobenzene-exposed workers and record-keeping of workplace exposure are needed.

12. FURTHER RESEARCH

12.1 Environmental issues

12.1.1 Exposure

For accuracy of exposure assessment modelling, more data are needed on the fate of nitrobenzene in soil, both in the root zone where plants are exposed and in the saturated and unsaturated zones where groundwater may become contaminated. Metabolism in plants is poorly characterized to date, so information on the nature and quantity of plant metabolites would assist assessment of exposure via that route. Similarly, more information about accumulation in plant tissues would be helpful.

12.1.2 Toxicity

More data are required on the chronic toxicity of nitrobenzene to aquatic organisms. Additional acute data are also required for marine and terrestrial species. Given that there is some doubt about the particular sensitivity of rainbow trout (*Oncorhynchus mykiss*) embryos and larvae to nitrobenzene, these studies should be repeated with appropriate quality assurance. Studies on the mechanism of nitrobenzene toxicity and any ameliorating environmental factors are also needed.

12.2 Health effects

There is limited accurate information on human exposure to nitrobenzene. More data on human exposure would be useful, estimated both from the measurement of environmental levels (especially air) and from biomarkers of exposure (e.g., *p*-nitrophenol in urine and haemoglobin adducts in the blood).

There are no data on the potential reproductive effects in humans exposed to nitrobenzene via any route. Data in animals showing obvious testicular toxicity would suggest that it would be worthwhile to explore this end-point among men exposed to nitrobenzene in the workplace or at hazardous waste sites. While it is known that nitrobenzene is a direct toxicant to the testis, information about its toxicity to male reproductive organs and subsequent developmental effects is insufficient and should be further studied.

Epidemiological studies on the working population should be conducted with a view to investigating effects on, for example, methaemoglobin levels, reproductive function, immunological status and neurobehavioural function.

Experimental findings in rats indicate that, although bacterial reduction in the gut is the primary *in vivo* metabolic step for nitrobenzene, there is some capacity for systemic reductive and oxidative metabolism. Nevertheless, germ-free rats are reported not to develop methaemoglobinaemia when dosed (intraperitoneally) with nitrobenzene. Since nitrobenzene causes methaemoglobinaemia in normal animals by the oral, dermal, inhalational and intraperitoneal routes (and humans by the oral, dermal and inhalational routes), this result could indicate that there may be some enterohepatic recycling of absorbed nitrobenzene, regardless of the route of exposure, thus allowing its access to gut bacteria. More detailed investigations of nitrobenzene toxicokinetics and metabolism should aid in the interpretation of the currently available animal studies and their relevance to humans and be helpful in making comparisons of human sensitivity with that of other animals.

The mode of action of nitrobenzene and its metabolites with respect to reproductive toxicity and carcinogenicity should be further studied.

13. PREVIOUS EVALUATIONS BY INTERNATIONAL BODIES

Nitrobenzene was reviewed by the International Agency for Research on Cancer (IARC) in 1996 (IARC, 1996). It was concluded in the evaluation that there was inadequate evidence in humans for the carcinogenicity of nitrobenzene, while there was sufficient evidence for carcinogenicity in experimental animals. Nitrobenzene was classified as possibly carcinogenic to humans (group 2B).

APPENDIX 1: THE CARCINOGENICITY OF INHALED NITROBENZENE — QUANTITATIVE RISK ASSESSMENTS

Specific dose–response considerations with respect to carcinogenic effects are presented in this appendix; several quantitative risk estimates are calculated and compared. Note that the quantitative risk estimate calculations presented in this appendix have been performed using methods generally employed by the US EPA, and none of these methods has been adopted by IPCS.

Quantitative risk assessment of inhaled nitrobenzene

In the only carcinogenicity studies on nitrobenzene conducted to date, male and female B6C3F$_1$ mice (70 per sex per group) were inhalationally exposed to 0, 25, 130 or 250 mg nitrobenzene/m^3 (0, 5, 25 or 50 ppm),[1] while male and female inbred Fischer-344 rats (60 per sex per group, plus 10 per sex intended for sacrifice after 15 months) and male Sprague-Dawley (CD) rats (60 per group) were exposed to 0, 5, 25 or 130 mg nitrobenzene/m^3 (0, 1, 5 or 25 ppm); all exposures were for 6 h per day, 5 days per week (excluding holidays), for a total of 505 days over 2 years (see section 7.3).

There is the mechanistic suggestion that the eight rodent organ responses observed may predict analogous cancer responses in humans similarly exposed to nitrobenzene (Holder, 1999a, 1999b). Hence, a quantitative assessment of *potential human risk* is appropriate for those exposures where nitrobenzene may affect humans (Holder, 1999b). This nitrobenzene quantitative risk assessment was based on the concerns raised from the positive cancer responses seen in the rodent bioassays (CIIT, 1993; Cattley et al., 1994). It proceeds on the basis that the number of animals per group equals 60, which the original protocol specified, while noting — qualitatively — that the extra

---

[1] In the calculations in this appendix, the conversion from ppm to mg/m^3 is performed using the actual temperature for the studies, i.e., 22.2 °C, that is, using a conversion factor mg/m^3 = 5.08 × ppm.

animals (10 per sex per group) had no tumours develop over the 2 years (Holder, 1999b).

If the true biologically based dose–response is unknown, current quantitative cancer risk assessment methods can only approximate the upper limit of expected risk; consequently, a conservative analysis often has to be made so as to cover the many unknown and partially understood variables. If a target-specific mechanism is not known, one approximation is to add affected cancer target sites in the most susceptible test animals. In the CIIT study, the male Fischer-344 rat malignant cancer incidences in the liver, thyroid and kidney are added to estimate the upper limit on potential cancer risk (Holder, 1999b). The cancer occurrences at 0, 5, 25 and 130 mg/m^3 in Fischer-344 male rats are, respectively, as follows: liver 1, 4, 5 and 16; thyroid 2, 1, 5 and 8; and kidney 0, 0, 0 and 6. Using the $N_{at\ risk} = 60$ (the number at risk originally set by CIIT), the combined respective incidences are 3/60 (0.050), 5/60 (0.083), 10/60 (0.167) and 26/60 (0.433). No time-to-tumour correction needs to be made because there was no differential mortality or significant early deaths. The statistical pairwise comparisons of each treated group combined incidence with the concurrent control group incidence yield probabilities (P) of 0.36, 0.037 and 3.4×10^{-7}. Although not actually statistically increased, the 5 mg/m^3 group incidence is arithmetically increased from 3/60 to 5/60, which suggests a possible fit into the trend set of carcinogenicity (Haseman & Lockhart, 1994).

The incidence of the observed combined concurrent controls, 0.05, and the combined historical control average incidence, 0.031 (i.e., weight-averaged among the three organs from past bioassays), are comparable. This suggests that the current experiment is responding as past bioassays have responded. The experimental doses 0, 5, 25 and 130 mg/m^3 can be converted into approximate human equivalent exposures of 0, 0.26, 1.3 and 6.48 mg/kg of body weight per day, which correlate with extra risks Π = 0, 0.0347, 0.123 and 0.403. In Figure A1, the human equivalent exposures are plotted as points against the corresponding extra risks on the y-axis. The best-fit curve (a coefficient of determination = 0.999) through the experimental cancer incidence points appears *supra*linear in curvature and second order in function. When an accepted pharmacokinetic mechanistic model is not available, an upper limit on risk can be set by the

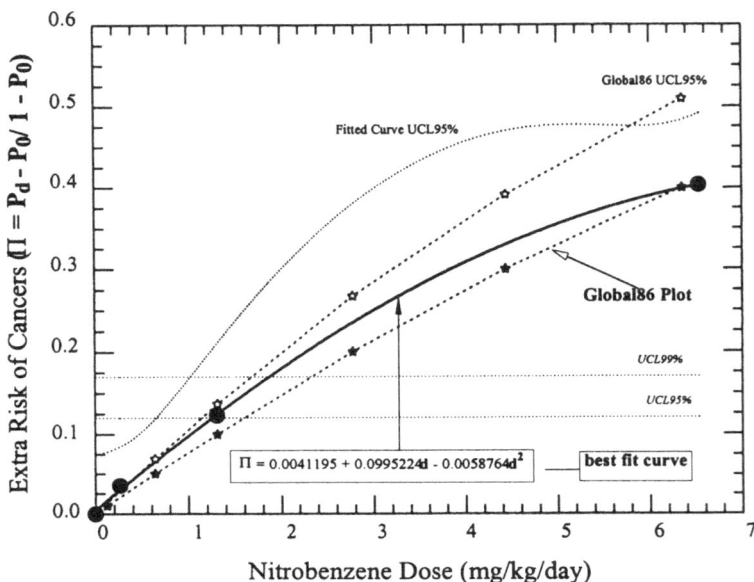

Fig. A1. Combined dose–response of inhaled nitrobenzene producing liver, thyroid and kidney cancers in male Fischer-344 rats.

y-axis: Probability of extra cancer occurrence or risk (abbreviated Π). Π is estimated by correcting for concurrent control rates P_0 as follows: $\Pi = (P_d - P_0)/(1 - P_0)$, and for each of the four rat exposures of 0, 1, 5 and 25 ppm, Π is, respectively, 0, 0.035, 0.1228 and 0.4035.

x-axis: Human equivalent nitrobenzene exposures (mg/kg of body weight per day).

Solid circles: The estimated human non-parametric, combined liver, thyroid and kidney carcinogenic responses of 0, 0.26, 1.31 and 6.53 mg/kg of body weight per day (2-year CIIT study).

Solid line: The best fit of the *estimated* non-parametric human responses is a second-order polynomial, $\Pi = 0.0041195 + 0.0995224d - 0.005876d^2$. The dose versus cancer correlation is 0.9997. A power function fits about as well, $\Pi = a \cdot d^b$, where $a = 0.092235$ and $b = 0.744979$.

Horizontal dashed lines: Indicate the statistical significance levels. Definite carcinogenicity is indicated in the 5 and 25 ppm nitrobenzene groups, but the 1 ppm response also seems to be part of the statistical trend. H_0 of no trend for the four exposures, $P < 10^{-8}$.

Upper dotted line: The upper 95% confidence limit (95%UCL) of the best fit, which is presented by the solid line through the points.

Dashed lines: The linearized multistage (LMS) model of Crump (1996) using Global86 software is plotted as the maximum likely estimate curve (lower dashed line, *filled stars*) and the 95%UCL curve (upper dashed line, *hollow stars*).

194

linearized multistage (LMS) model (Crump, 1996). This is intended to be only an interim method of risk estimation until the mode of action becomes known. This LMS model is basically a polynomial fitting of the bioassay data, maintaining the constraints that 1) the experimental curve is fitted as well as possible, 2) the lowest fitted point in the experimental range is connected with zero-dose controls and 3) the assumption of *possible* positive cancer responses at all doses >0 (Crump et al., 1977; E.L. Anderson, 1983; Crump, 1996). The LMS model, as usually employed, estimates the upper 95% confidence limit (95%UCL) on the best-fitted curve, and this 95%UCL curvature describes a reasonably stable dose–response curve over many orders of dose. It is a limit curve.

The LMS model fit is depicted in Figure A1 and only approximates the best-fit curve of the points in the experimental range because of the following constraints in the Global86 program.[1] The LMS model 1) censures negative coefficient terms (compare with the equation in Figure A1), 2) forces the presence of the αd^1 linear term (hence, the *linear* multistage model) and 3) is optimized for lower dose group responses and the expectation of low-dose, non-zero, non-negative Π values down through the environmental exposure level estimates of the 95%UCL on the fitted curve (Crump et al., 1977; Crump, 1996; Holder, 1999b).

The Global86 maximum likely estimate (or the best fit) of the cancer slope is 0.077/mg per kg of body weight per day, whereas the 95%UCL curve slope (called the q_1*) = 0.11 per mg per kg of body weight per day (Holder, 1999b). The q_1* of the 95%UCL curve is the more stable and conservative estimator of the cancer slope at low exposures (E.L. Anderson, 1983). The cancer risk, at low exposures, is assumed to be simply estimated as the product of the q_1* and the anticipated nitrobenzene average exposure in mg/kg of body weight per day. The use of q_1* is meant to be only an approximation of a risk

---

[1] Global86 is a statistical program that uses a technique where the upper bound at 95% (95%UCL) of a set of cancer dose–response data is estimated by optimization methods (Guess & Crump, 1976). This technique is used to extrapolate from the data fit range to lower exposures, with the assumption that the true potency (slope) derived from the 95%UCL will not likely be higher than the UCL at any exposure of concern and could be less, or even zero potency (Crump, 1996).

limit (95%UCL curve), and the true value of the cancer slope could be less, or even zero (Albert, 1994).

As a regulatory reference, the virtually safe dose (VSD) is that exposure that may give rise to an Π of 1 in 1 000 000, a rare event (Gaylor, 1989). From the LMS model, the VSD can be estimated to be the chosen risk / q_1* (Gaylor & Gold, 1995). So, VSD = $10^{-6}/0.11$ per mg per kg of body weight per day = 0.91×10^{-5} mg/kg of body weight per day or 9.1 ng/kg of body weight per day. A liver-only estimation of VSD is 15.6 ng/kg of body weight per day. Considering margin of error with these estimates, this suggests that adding the three tissues as if one target does not significantly affect the VSD estimate (9.1 versus 15.6 ng/kg of body weight per day).

Comparison of the quantitative estimate with other methods

To place the above LMS estimate in perspective, a number of other previously used cancer risk models have been examined and compared with the LMS model (Holder, 1999b). These models are presented below in Table A1. The LMS model will be discussed first.

In the US EPA, the standard approach to chemical carcinogen risk assessment, when the mechanism is generally not known, is to employ the LMS model, in which it is assumed that there is 1) a linear dose–response in the low-dose region and 2) no practical threshold (Guess & Crump, 1976; E.L. Anderson, 1983; Albert, 1994). The concurrent control rates are high enough that the assumption of linearity in the low-dose range is supported (cf. Gaylor, 1992). The estimation of the unit upper limit cancer slope for nitrobenzene, q_1*, is 0.11 per mg per kg of body weight per day, and the VSD is estimated to be 9.1 ng/kg of body weight per day of continuous exposure.

One of the earliest analyses of cancer dose–response suggested using "scientific judgement" to quantify carcinogenic risk (Weil, 1972). Weil suggested that any scheme in which carcinogen-related multiple events are employed to produce tumours is not consistent with a linear function and that "apparent" thresholds do occur in many bioassay analyses. Because there is so much uncertainty due to experimental variability and often replication is not done, many "cancer potency slopes" might arise in practice. Which particular slope applies to human exposure situations is problematic according to Weil (1972). He suggested a simple factoring down of the lowest dose producing cancer to a reasonably safe dose achieved by best technology. He suggested a factor of 1/5000. Hence, using this recommendation, the lowest nitrobenzene dose presenting an increase is 1.3 mg/kg of body weight per day, which can be subjectively factored down by 1/5000 to a dose of 260 ng/kg of body weight per day.

Table A1. Comparison of virtually safe dose (VSD)[a] estimates using different methods

Method	VSD[b] (ng/kg of body weight per day)
Gaylor	8.8
LMS[c]	9.1
BMD[d]	9.4
1-Hit	13.3
Liver[e]	15.6
Zeise	52.6
BMD/safety factor	94
Weil	260

[a] Comparison is made of various risk methods that have been used, are currently being used or are being considered for use. The VSDs are arranged in increasing order of exposure magnitude from top to bottom, i.e., less conservative estimates are at the bottom. Because the VSD is an exposure, it is presented in an exposure unit (as ng/kg of body weight per day). The VSD may be considered a reference level for a lifetime of nitrobenzene exposure.

[b] A VSD may be assumed to be a risk of 1 in a million. The VSD is an arbitrary exposure and not necessarily an absolute level of safety. The VSD is an exposure correlated to very low risk that may be considered as a convenient reference exposure level when considering other environmental exposures.

[c] Linearized multistage (LMS) method upper confidence limit at the 95% level of risk. Estimated from the sum of the liver, thyroid and kidney responses using the Global86 computer program. In the absence of a specific biologically driven dose–response model, this method is used currently in the USA.

[d] Benchmark dose, a method that assumes continuity and no demonstrable threshold for the cancer response.

[e] Liver response only was modelled by the LMS method using one organ response, not three organ sites (see note c).

The one-hit model is stated as $\Pi = 1 - e^{-\beta \cdot d}$, where β is the slope and the "$\beta \cdot d$" term is the number of hits of a single type (Klaassen, 2001). The one-hit model estimates the nitrobenzene cancer slope as $\beta = 0.075$ per mg per kg of body weight per day, and the VSD = 13.3 ng/kg of body weight per day (Table A1). Another past model to compare with the LMS is based on the observation that for many animal carcinogens, the cancer potency seems linked empirically to the lethal toxicity, or LD_{50} or LC_{50} (Zeise et al., 1984, 1986). In this empirical method, the LC_{50} for nitrobenzene is approximately 183 mg/kg of body weight per day, from which the slope is empirically estimated to be 0.019 per mg per kg of body weight per day (Zeise et al., 1984, 1986). The slope in turn leads to: 1/slope = VSD = 52.6 ng/kg of body weight per day.

Yet another method approximates the VSD from the maximum tolerated dose (MTD), usually determined in a 90-day subchronic test (Gaylor, 1989; Krewski et al., 1993). The MTD is often the same as, or near, the maximum dose tested (MDT), which for nitrobenzene is 130 mg/m^3, or 6.48 mg/kg of body weight per day. It has been determined for many chemicals tested so far that the MTD/VSD is approximately a constant 7.40×10^5 (unitless), with about a ±10-fold variation (Gaylor & Gold, 1995). Hence, it was suggested that the VSD may be estimated by merely dividing the MDT by 7.40×10^5. For nitrobenzene, VSD = (6.48 mg/kg of body weight per day)/(7.40×10^5) = 8.8 ng/kg of body weight per day (Table A1; Holder, 1999b). Parenthetically, the ratio MDT/VSD for nitrobenzene (from the LMS model) is 7.12 and compares well with the geometric average value of 7.4×10^{-5} for the 324 chemicals studied by Gaylor (1989).

A benchmark dose (BMD) method of estimating cancer risks has been proposed (Crump, 1984; US EPA, 1996; Gaylor & Gold, 1998). This method determines a point on the experimental toxicity curve in the lower region of response called a point of departure (Gaylor & Gold, 1998). This point often is taken at $\Pi = 0.10$ and is connected to the zero–zero origin, thereby making a straight line. The risk at any lower environmental dose is determined as coordinates from this straight "upper bound" line (US EPA, 1996). The BMD for nitrobenzene using this method is determined to be 0.94 mg/kg of body weight per day; assuming no threshold and likely direct genetic effects (method 1), the VSD at $1/10^6$ risk is estimated to be 9.4 ng/kg of body weight per day. A safety factor might be constructed if a threshold were to be demonstrated (method 2). By assuming, for example, a 10-fold variability in each of the factors such as animal-to-human and intrahuman variability, variability of low-dose slope (tailing), and irreversibility and severity of cancer, then the BMD may be divided by

10^4 to get a reasoned safety estimate of 940 ng/kg of body weight per day, or a VSD of 94 ng/kg of body weight per day (Table A1).

Discussion of the quantitative estimates

Nitrobenzene inhalation exposure causes cancer in eight sites, three of which are malignant responses, in three species: $B6C3F_1$ mice (males and females), Fischer-344 rats (males and females) and male Sprague-Dawley rats (females not tested). This distributed cancer response among organs, sex and species of rodents has been associated qualitatively with the metabolic formation of nitrophenols, nitroxides, nitro free radicals and superoxide free radicals (Holder, 1999a). The carcinogenic potency of all these toxicants is unknown. The curvature of the combined cancer response curve for nitrobenzene in male Fischer-344 rats appears *supra*linear (convex) and does not suggest a threshold (Figure A1). However, some of the low nitrobenzene back-ground tumour responses might exhibit a threshold if they were analysed individually and in bioassays of increased sensitivity (i.e., with more test animals). Although no experimental data exist on the low-dose continuity of the dose–response curve, further studies on the mechanism of action of nitrobenzene should examine this aspect of the response. The various VSD estimates from these models (Table A1) vary over a 30-fold range. Such a spread is not at all an extreme variance, considering the varied methods and assumptions made in those models. This is likely due to the prominence of the linear com-ponent in the nitrobenzene response data and the fact that some of the VSD estimator methods are interrelated. However, it also suggests that any dependable method of the group does not seem to disagree with any of the other methods, albeit based on different assumptions.

In conclusion, until a true pharmacodynamic and pharmacokinetic model becomes available for nitrobenzene, the $q_1^* = 0.11$ per mg per kg of body weight per day slope may serve as a public protective measure in estimating human exposure upper limit risks from nitro-benzene environmental exposures. It is quite likely that biologically based modelling using critical, actual dose–response parameters will demonstrate a higher VSD than the estimated VSD of 9.1 ng/kg of body weight per day here. This assumption is based on the fact that the methods presented here (except the Weil [1972] method) are inher-ently conservative.

APPENDIX 2:
OUTLINE OF THE DUTCH STATISTICAL EXTRAPOLATION METHOD USED TO DERIVE A GUIDELINE VALUE FOR NITROBENZENE FOR THE PROTECTION OF AQUATIC SPECIES

Introduction

The traditional approach to using single-species toxicity data to protect field ecosystems has been to apply arbitrary *assessment factors*, *safety factors* or *application factors* to the lowest toxicity figure for a particular chemical. The magnitude of these safety factors depends on whether acute or chronic toxicity figures are available and the degree of confidence that one has in whether the figures reflect the field situation. Most of the factors used are multiples of 10, and larger factors are applied where there is less certainty in the data. For example, a factor of 1000 is generally used for acute data except for essential elements, in which case a factor of 200 is applied. This factor of 200 includes a factor of 10 for extrapolating from laboratory to field, a further factor of 10 for a limited data set and a factor of 2 for conversion of an acute end-point to a chronic end-point (e.g., for an essential metal).

Concerns have often been raised as to the arbitrary nature of assessment factors (Chapman et al., 1998) and the fact that they do not conform to risk assessment principles. The OECD (1992) recommended that assessment factors be used only when there are inadequate data for statistical extrapolation methods to be used.

The following sections briefly outline the statistical extrapolation method used to derive the nitrobenzene guideline for protection of aquatic organisms (see chapter 10). Much of the text is taken directly from the Australian and New Zealand Guidelines for Fresh and Marine Water Quality (ANZECC/ARMCANZ, 2000).

Use of statistical extrapolation methods

New methods using statistical risk-based approaches have been developed over the last decade for deriving guideline (trigger) values. These are based on calculations of a statistical distribution of laboratory ecotoxicity data and attempt to offer a predetermined level of protection, usually 95%. The approach of Aldenberg & Slob (1993) has been adopted in the Netherlands, Australia and New Zealand for guideline derivation and is recommended for use by the OECD. It was chosen because of its theoretical basis, its ease of use and the fact that it has been extensively evaluated. Warne (1998) has compared in detail the risk-based and assessment factor approaches used in various countries.

The Aldenberg & Slob (1993) method uses a statistical approach to protect 95% of species with a predetermined level of confidence, provided there is an adequate data set. This approach uses available data from all tested species (not just the most sensitive species) and considers these data to be a subsample of the range of concentrations at which effects would occur in all species in the environment. The method may be applied if toxicity data, usually chronic NOEC values, are available for at least five different species from at least four taxonomic groups. Data are entered into a computer program EcoToX (ETX) (Aldenberg, 1993) and generally fitted to a log-logistic distribution. A hazardous concentration for p per cent of the species (HC_p) is derived. HC_p is a value such that the probability of selecting a species from the community with a NOEC smaller than HC_p is equal to p (e.g., 5%, HC_5). HC_5 is the estimated concentration that should protect 95% of species. A level of uncertainty is associated with this derived value, and so values with a given confidence level (e.g., 50% or 95%) are computed in the ETX program by attaching a distribution to the error in the tail (Figure A2). The ANZECC/ARMCANZ (2000) guidelines use the median of 50% confidence.

HC_5 (or the 95% protection level) is estimated using the ETX approach by dividing the geometric mean of the NOEC values for m species by an extrapolation factor K (OECD, 1995), where:

$$K = \exp^{(s_m \times k)}$$

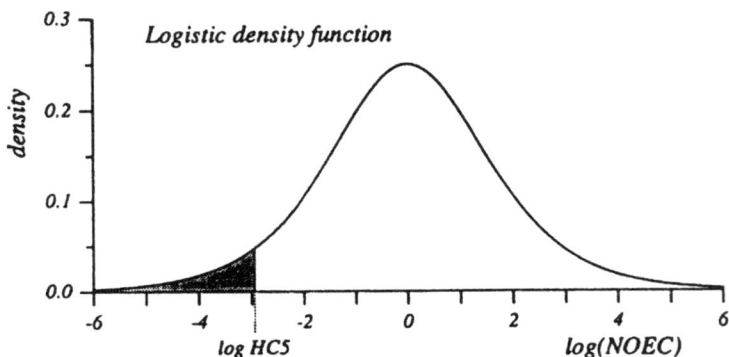

Fig. A2. The Dutch statistical approach for the derivation of trigger values (from Aldenberg & Slob, 1993).

and:

s_m =   sample standard deviation of natural logarithm of the NOEC values for m species;

k =   one-sided tolerance limit factor for a logistic or normal distribution (from computer simulations).

Where acute LC_{50} data are used to derive a *moderate reliability* trigger value, the figure resulting from the statistical distribution model is converted to a chronic trigger value, using an acute-to-chronic (LC_{50}-to-NOEC) conversion.

The Aldenberg & Slob (1993) extrapolation method is based on several critical assumptions, outlined below. Many of these are common to other statistical distribution methods:

- The ecosystem is sufficiently protected if theoretically 95% of the species in the system are fully protected.

- The distribution of the NOECs is symmetrical.

- The available data are derived from independent random trials of the total distribution of sensitivities in the ecosystem.

- Toxicity data are distributed log-logistically, i.e., a logistic distribution is the most appropriate to use.

- There are no interactions between species in the ecosystem.

- NOEC data are the most appropriate data to use to set ambient environmental guidelines.

- NOEC data for five species are a sufficient data set.

Modification of the Aldenberg and Slob approach

The Aldenberg & Slob (1993) approach assumes that the data are best fitted to a log-logistic distribution. For some data sets, however, a better fit is obtained with other models. By using a program developed by CSIRO Biometrics, the data are compared with a range of statistical distributions called the Burr family of distributions, of which the log-logistic distribution is one case. The program determines the distribution that best fits the available toxicity data and calculates the 95% protection level with 50% confidence (ANZECC/ ARMCANZ, 2000). This method has been used to calculate the HC_5 for nitrobenzene below.

Derivation of the guideline trigger value for nitrobenzene

Acute LC_{50} values used in the freshwater trigger value derivation included the following:

- fish: six species with geometric mean 48- to 96-h LC_{50} values ranging from 1.8 to 138 mg/litre
- crustaceans: two species, *Daphnia magna* (with a geometric mean 24- to 48-h LC_{50} of 36 mg/litre) and *Ceriodaphnia dubia* (with a 24-h LC_{50} of 54 mg/litre)
- protozoans: one species with a 24-h EC_{50} of 98 mg/litre
- algae: three species with 96-h EC_{50}s ranging from 17.8 to 43 mg/litre
- bacteria: one species with a geometric mean 15-min EC_{50} of 23 mg/litre
- amphibians: one species with 48-h LC_{50} of 121 mg/litre
- other invertebrates, including a flatworm (168-h LC_{50} of 2 mg/litre), snail (geometric mean 48-h LC_{50} of 95 mg/litre) and midge (48-h LC_{50} of 70 mg/litre)

The acute trigger value for nitrobenzene derived in the program from these data was 3230 µg/litre. However, this has to be converted to a chronic value. The only chronic toxicity data available were for *Daphnia magna*, with NOECs of 2.6 and 1.9 mg/litre (geometric mean NOEC of 2.2 mg/litre). The geometric mean acute value for this species (36 mg/litre) was divided by the chronic value (2.2 mg/litre) to give an acute:chronic ratio of 16. Applying this acute:chronic ratio to the acute trigger value:

$$3230/16 \approx 200 \text{ µg/litre}$$

Thus, the concentration limit for nitrobenzene, to protect 95% of species with 50% confidence, is about 200 µg/litre.

Data for marine species were limited, so no trigger value could be derived using this method.

Conclusions

In keeping with the recent development of risk-based statistical approaches to derive guidelines for the protection of aquatic organisms, a freshwater guideline value for nitrobenzene to protect 95% of species with 50% confidence was derived using a modified method of Aldenberg & Slob (1993). The derived value was 200 µg/litre (see chapter 10).

REFERENCES

Abbinante R & Pasqualato D (1997) Intoxication due to ingestion of bitter almond oil contaminated with nitrobenzene. Toxicologist, **36**(1), Part 2, Poster 218.

Abe S & Sasaki M (1977) Chromosome aberrations and sister chromatid exchanges in Chinese hamster cells exposed to various chemicals. J Natl Cancer Inst, **58**: 1635–1641.

ACGIH (1991) Documentation of the threshold limit values and biological exposure indices, 6th ed. Cincinnati, Ohio, American Conference of Governmental Industrial Hygienists, vol 2, pp 1096–1099.

ACGIH (1995) 1995–1996 threshold limit values (TLVs) for chemical substances and physical agents and biological exposure indices (BEIs). Cincinnati, Ohio, American Conference of Governmental Industrial Hygienists, pp 28, 65.

ACGIH (2000) 2000 TLVs® and BEIs®. Cincinnati, Ohio, American Conference of Governmental Industrial Hygienists, p 100.

Acosta de Pérez O, Bernacchi AS, Diaz de Toranzo EG, & Castro JA (1992) Reductive biotransformation of xenobiotics by the sheep ruminal content. Comp Biochem Physiol C, **101**: 625–626.

Adkins RL (1996) Nitrobenzene and nitrotoluenes. In: Kroschwitz JI & Howe-Grant M ed. Kirk-Othmer encyclopedia of chemical technology, 4th ed. New York, John Wiley & Sons, vol 17, pp 133–152.

Adler A (1934) Pallidar syndrome with hyperkinesis and "Zwangsdenken" as sequelae of nitro-benzene poisoning. Z Gesamte Neurol Psychiat, **150**: 341–345 [cited in Von Oettingen, 1941].

Adler B, Braun R, Schoeneich J, & Boehme H (1976) Repair-defective mutants of *Proteus mirabilis* as a prescreening system for the detection of potential carcinogens. Biol Zentralbl, **95**: 463 [cited in Beauchamp et al., 1982].

Albert RE (1994) Carcinogen risk assessment in the U.S. Environment Protection Agency. Crit Rev Toxicol, **24**: 75–85.

Albrecht W & Neumann H-G (1985) Biomonitoring of aniline and nitrobenzene: Hemoglobin binding in rats and analysis of adducts. Arch Toxicol, **57**: 1–5.

Aldenberg T (1993) ETX 1.3a. A program to calculate confidence limits for hazardous concentrations based on small samples of toxicity data. Bilthoven, National Institute of Public Health and Environmental Protection (No. 719102015).

Aldenberg T & Slob W (1993) Confidence limits for hazardous concentrations based on logistically distributed NOEC toxicity data. Ecotoxicol Environ Saf, **25**: 48–63.

Alexander M & Lustigman BK (1966) Effect of chemical structure on microbial degradation of substituted benzenes. J Agric Food Chem, **14**: 410–413.

Allenby G, Sharpe RM, & Foster PM (1990) Changes in Sertoli cell function *in vitro* induced by nitrobenzene. Fundam Appl Toxicol, **14**: 364–375.

Allenby G, Foster PM, & Sharpe RM (1991) Evaluation of changes in the secretion of immunoactive inhibin by adult rat seminiferous tubules *in vitro* as an indicator of early toxicant action on spermatogenesis. Fundam Appl Toxicol, **16**: 710–724.

Altschuh J, Brüggemann R, Santl H, Eichinger G, & Piringer OG (1999) Henry's law constants for a diverse set of organic chemicals: Experimental determination and comparison of estimation methods. Chemosphere, **39**: 1871–1887.

Amacher DE, Paillet SC, Turner GN, Ray VA, & Salsburg DS (1980) Point mutations at the thymidine kinase locus in L5178Y mouse lymphoma cells. II. Test validation and interpretation. Mutat Res, **72**: 447–474.

Amoore JE & Hautala E (1983) Odor as an aid to chemical safety: Odor thresholds compared with threshold limit values and volatilities for 214 industrial chemicals in air and water dilution. J Appl Toxicol, **3**: 272–290.

Anbar M & Neta P (1967) A compilation of specific bimolecular rate constants for the reactions of hydrated electrons, hydrogen atoms and hydroxyl radicals with inorganic and organic compounds in aqueous solution. Int J Appl Radiat Isot, **18**: 493–523.

Anderson D & Styles JA (1978) The bacterial mutation test. Br J Cancer, **37**: 924–930.

Anderson EL (1983) Quantitative approaches in use to assess cancer risk. Risk Anal, **3**: 277–295.

Anderson GE (1983) Human exposure to atmospheric concentrations of selected chemicals. Vol II. Research Triangle Park, North Carolina, US Environmental Protection Agency, Office of Air Quality Planning and Standards (NTIS No. PB83-265249).

Anderson TA, Beauchamp JJ, & Walton BT (1991) Organic chemicals in the environment. Fate of volatile and semivolatile organic chemicals in soils: abiotic versus biotic losses. J Environ Qual, **20**: 420–424.

Anonymous (1987) Chemical profiles: Nitrobenzene. Chemical Marketing Reporter, 3 August, p 50.

ANZECC/ARMCANZ (2000) Australian and New Zealand guidelines for fresh and marine water quality. National Water Quality Management Strategy, Australian and New Zealand Environment Conservation Council and Agriculture and Resource Management Council of Australia and New Zealand (further information can be found at http://www.environment.gov.au/water/quality/).

Arinç E & Aydoğmus A (1990) Lung microsomal *p*-nitrophenol hydroxylase — characterization and reconstitution of its activity. Comp Biochem Physiol B, **97**: 455–460.

Arnts RR, Seila RL, & Bufalini JJ (1987) Determination of OH rate constants for six volatile organics in air using the UCR protocol. Washington, DC, US Environmental Protection Agency (Report No. EPA-600/D-87-197).

Aßmann N, Emmrich M, Kampf G, & Kaiser M (1997) Genotoxic activity of important nitrobenzenes and nitroanilines in the Ames test and their structure–activity relationship. Mutat Res, **395**: 139–144.

Astier A (1992) Simultaneous high-performance liquid chromatographic determination of urinary metabolites of benzene, nitrobenzene, toluene, xylene and styrene. J Chromatogr, **573**: 318–322.

Atkinson R (1985) Kinetics and mechanisms of the gas-phase reactions of the hydroxyl radical with organic compounds under atmospheric conditions. Chem Rev, **85**: 69–201.

Atkinson R (1990) Lifetimes and fates of toxic air contaminants in California's atmosphere. Riverside, California, University of California, Statewide Air Pollution Research Centre (Report ARB-R-90/441).

Atkinson R, Tuazon EC, Wallington TJ, Aschmann SM, Arey J, Winer AM, & Pitts JN Jr (1987) Atmospheric chemistry of aniline, *N,N*-dimethylaniline, pyridine, 1,3,5-triazine and nitrobenzene. Environ Sci Technol, **21**: 64–72.

ATSDR (1990) Toxicological profile for nitrobenzene. Atlanta, Georgia, US Department of Health and Human Services, Public Health Service, Agency for Toxic Substances and Disease Registry (Publication No. TP-90-19).

Bader M, Goen T, Muller J, & Angerer J (1998) Analysis of nitroaromatic compounds in urine by gas chromatography–mass spectrometry for the biological monitoring of explosives. J Chromatogr B, Biomed Sci Appl, **710**(1–2): 91–99.

Bahnick DA & Doucette WJ (1988) Use of molecular connectivity indices to estimate soil sorption coefficients for organic chemicals. Chemosphere, **17**: 1703–1715.

Baider LM & Isichenko TS (1990) [EPR studies and the role of the electron structure of the heterocycle in the metabolic denitrosation of 5-nitrofuran.] Izv Akad Nauk SSSR Biol, **6**: 925–928 (in Russian).

Bandow H, Washida N, & Akimoto H (1985) Ring-cleavage reactions of aromatic hydrocarbons studied by FT-IR spectroscopy. I. Photooxidation of toluene and benzene in the NO_x–air system. Chem Soc Jpn, **58**: 2531–2540.

Banerjee S, Yalkowsky SH, & Valvani SC (1980) Water solubility and octanol/water partition coefficients of organics. Limitation of the solubility–partition coefficient correlation. Environ Sci Technol, **14**: 1227–1229.

Bangert S, McDonald DB, Prior SJ Jr, & Tuthill SJ (1988) Manufactured gas waste disposal investigations. Two case studies in Iowa. In: Proceedings of the 42nd Industrial Waste Conference, 12–14 May 1987. Boca Raton, Florida, Lewis Publishers, pp 39–51.

Barth EF & Bunch RL (1979) Removability, biodegradation and treatability of specific pollutants. Washington, DC, US Environmental Protection Agency (Report No. EPA-600/9-79-034; NTIS No. PB80-106-438).

Bartsch H, Malaveille C, Camus AM, Martel-Planche G, Brun G, Hautefeuille A, Sabadie N, Barbin A, Kuroki T, Drevon C, Piccoli C, & Montesano R (1980) Validation and comparative studies on 180 chemicals with *S. typhimurium* strains and V79 Chinese hamster cells in the presence of various metabolizing systems. Mutat Res, **76**: 1–50.

BASF (1977) Bericht über die Prüfung der akuten Inhalationsgefahr von Nitrobenzol an der Ratte. Ludwigshafen, BASF Aktiengesellschaft (unpublished).

BASF (1997) Investigation of the potential for DNA binding of nitrobenzene. Ludwigshafen, BASF Aktiengesellschaft (unpublished).

BASF, Bayer, & Zeneca (1995) Cytogenic study *in vivo* of nitrobenzene in mice, micronucleus test, single intraperitoneal administration (unpublished report).

Beard RR & Noe JT (1981) Aromatic nitro and amino compounds. In: Clayton GD & Clayton FE ed. Patty's industrial hygiene and toxicology, 3rd ed. Vol 2A: Toxicology. New York, John Wiley & Sons, pp 2416–2417, 2430–2434, 2469, 2484–2489.

Beauchamp ROJ, Irons RD, Rickert DE, Couch DB, & Hamm TEJ (1982) A critical review of the literature on nitrobenzene toxicity. CRC Crit Rev Toxicol, **11**: 33–84.

Benkendorf C (1978) Metabolic pathways of the *N*-substituted naphthalenes: their relationship to mutagenicity and carcinogenicity. Diss Abstr Int, **B39**: 2723-B.

Berger J (1990) Bone marrow and spleen morphology following haemolytic anaemia in adult laboratory rats of various age groups. Folia Haematol Int Mag Klin Morphol Blutforsch, **117**: 1–5.

Billings RE (1985) Mechanisms of catechol formation from aromatic compounds in isolated rat hepatocytes. Drug Metab Dispos, **13**: 287–290.

Bio/dynamics Inc. (1983) A range-finding study to evaluate the toxicity of nitrobenzene in the pregnant rabbit. East Millstone, New Jersey, Bio/dynamics Inc. (Project No. 83-2723; unpublished).

Bio/dynamics Inc. (1984) An inhalation teratology study in rabbits with nitrobenzene. Conducted for the Nitrobenzene Association Toxicology Task Group. East Millstone, New Jersey, Bio/dynamics Inc., 16 August (Project No. 83-2725; unpublished).

Blaauboer BJ & Van Holsteijn CW (1983) Formation and disposition of *N*-hydroxylated metabolites of aniline and nitrobenzene by isolated rat hepatocytes. Xenobiotica, **13**: 295–302.

Black JA, Birge WJ, McDonnell WE, Westerman AG, Ramey BA, & Bruser DM (1982) The aquatic toxicity of organic compounds to embryo–larval stages of fish and amphibians. Lexington, Kentucky, University of Kentucky, Water Resources Research Institute (Research Report No. 133; NTIS No. PB82-224601).

Blum DJ & Speece RE (1991) A database of chemical toxicity to environmental bacteria and its use in interspecies comparisons and correlations. J Water Pollut Control Fed, **63**: 198–207.

Bollman MA, Wanda KB, Smith S, DeWhitt K, & Kapustka L (1990) Report on algal toxicity tests on selected Office of Toxic Substances (OTS) chemicals. Corvallis, Oregon, US

Environmental Protection Agency, pp 1–42 (Report No. EPA/600/3-90/041; NTIS No. PB-90-212606).

Bond JA, Chism JP, Rickert DE, & Popp JA (1981) Induction of hepatic and testicular lesions in Fischer-344 rats by single oral doses of nitrobenzene. Fundam Appl Toxicol, **1**: 389–394.

Borghoff SJ (1999) Male rat renal carcinogens: Distinguishing the contribution of the α2u globulin-mediated mechanism. CIIT Activ, **19**(10): 5–6.

Bozzelli JW & Kebbekus BB (1982) A study of some aromatic and halocarbon vapors in the ambient atmosphere of New Jersey. J Environ Sci Health, **A17**: 693–711.

Bozzelli JW, Kebbekus BB, & Greenberg A (1980) Analysis of selected toxic and carcinogenic substances in ambient air at New Jersey. Trenton, New Jersey, New Jersey Department of Environmental Protection, Office of Cancer and Toxic Substances Research.

Briggs GG (1973) Proceedings of the 7th British Insecticides and Fungicides Conference, p 83.

Briggs GG (1981) Theoretical and experimental relationships between soil adsorption, octanol–water partition coefficients, water solubilities, bioconcentration factors, and the parachor. J Agric Food Chem, **29**: 1050–1059.

Bringmann G & Kühn R (1959) Vergleichende wasser-toxicologische Untersuchungen an Bakterien, Algen und Kleinkrebsen. Gesundheits-Ingenieur, **4**: 115–120.

Bringmann G & Kühn R (1978) Grenzwerte der Schadwirkung wassergefahrdender Stoffe gegen Blaualgen (*Microcystis aeruginosa*) und Grunalgen (*Scenedesmus quadricauda*) im Zellvermehrungshemmtest. Vom Wasser, **80**: 45–60.

Bringmann G & Kühn R (1980) Comparison of the toxicity thresholds of water pollutants to bacteria, algae, and protozoa in the cell multiplication inhibition test. Water Res, **14**: 231–241.

Bringmann G & Kühn R (1982) Results of toxic action of water pollutants on *Daphnia magna* Straus tested by an improved standardized procedure. Z Wasser Abwasser Forsch, **15**: 1–6.

Bryant C & DeLuca M (1991) Purification and characterization of an oxygen-insensitive NAD(P)H nitroreductase from *Enterobacter cloacae*. J Biol Chem, **266**: 4119–4125.

BUA (1993) 2-Nitrophenol and 4-nitrophenol. GDCh-Advisory Committee on Existing Chemicals of Environmental Relevance. Stuttgart, S. Hirzel Wissenscaftliche Verlagsgesellschaft, 168 pp (English translation of BUA Report 75, published in 1992).

BUA (1994) Nitrobenzene. GDCh-Advisory Committee on Existing Chemicals of Environmental Relevance. Stuttgart, S. Hirzel Wissenscaftliche Verlagsgesellschaft, 100 pp (English translation of BUA Report 59, published in 1991).

BUA (1995) OH-radicals in the troposphere: Concentration and effects. GDCh-Advisory Committee on Existing Chemicals of Environmental Relevance. Stuttgart, S. Hirzel Wissenscaftliche Verlagsgesellschaft, 163 pp (English translation of BUA Report 100, published in 1993).

Buccafusco RJ, Ells SJ, & LeBlanc GA (1981) Acute toxicity of priority pollutants to bluegill (*Lepomis macrochirus*). Bull Environ Contam Toxicol, **26**: 446–452.

Burns & Roe (1982) Fate of priority pollutants in publicly owned treatment works. Vol 1. Paramas, New Jersey, Burns and Roe Industrial Services Organization; Washington, DC, US Environmental Protection Agency (Report No. EPA-440/1-82/303).

Burns LA, Bradley SG, White KL Jr, McCay JA, Fuchs BA, Stern M, Brown RD, Musgrove DL, Holsapple MP, Luster MI, & Munson AE (1994) Immunotoxicity of nitrobenzene in female B6C3F$_1$ mice. Drug Chem Toxicol, **17**: 271–315.

Burns LH, Cline DM, & Lassiter RR (1982) Exposure analysis model system (EXAMS). Athens, Georgia, US Environmental Protection Agency, Environmental Research Laboratory (Report No. EPA-600/3-82-023).

Bus JS (1983) Aniline and nitrobenzene: Erythrocyte and spleen toxicity. CIIT Activ, **3**(12): 1, 6.

Bus JS & Gibson JE (1982) Mechanisms of superoxide radical-mediated toxicity. J Toxicol Clin Toxicol, **19**(6–7): 689–697.

Butterworth BE, Smith-Oliver T, Earle L, Loury DJ, White RD, Doolittle DJ, Working PK, Cattley RC, Jirtle R, Michalopoulos G, & Strom S (1989) Use of primary cultures of human hepatocytes in toxicology studies. Cancer Res, **49**: 1075–1084.

Callahan MA, Slimak MW, Gabel NW, May IP, Fowler CF, Freed JR, Jennings P, Durfee RL, Whitmore FC, Maestri B, Mabey WR, Holt BR, & Gould C (1979) Water-related environmental fate of 129 priority pollutants. Vol II. Washington, DC, US Environmental Protection Agency, Office of Water Planning and Standards (Report No. EPA-440/4-79-029b; NTIS No. PB80-204381).

Camara E, Leder AE, & Ishikawa Y (1997) CEH data summary: Nitrobenzene. In: Chemical economics handbook. Menlo Park, California, SRI International.

Canton JH, Slooff W, Kool HJ, Struys J, Pouw TJM, Wegman RCC, & Piet GJ (1985) Toxicity, biodegradability and accumulation of a number of C1/N containing compounds for classification and establishing water quality criteria. Regul Toxicol Pharmacol, **5**: 123–131.

Carter FW (1936) An unusual case of poisoning, with some notes on non-alkaloidal organic substances. Med J Aust, **24**: 558–564.

Caspary WJ, Daston DS, Myhr BC, Mitchell AD, Rudd CJ, & Lee PS (1988) Evaluation of the L5178Y mouse lymphoma cell mutagenesis assay: Interlaboratory reproducibility and assessment. Environ Mol Mutagen, **12**(Suppl 13): 195–229.

Cattley RC, Everitt JI, Gross EA, Moss OR, Hamm TEJ, & Popp JA (1994) Carcinogenicity and toxicity of inhaled nitrobenzene in B6C3F$_1$ mice and F344 and CD rats. Fundam Appl Toxicol, **22**: 328–340.

Cerutti PA (1994) Oxy-radicals and cancer. Lancet, **344**: 862–863.

Chambers JV & O'Neill FJ (1945) Nitrobenzene poisoning; report of a fatal case. US Navy Med Bull (Washington), **44**: 1112–1115.

Chapman PM, Fairbrother A, & Brown D (1998) A critical evaluation of safety (uncertainty) factors for ecological risk assessment. Environ Toxicol Chem, **17**: 99–108.

Chapman WM & Fox CG (1945) Nitrobenzene poisoning from furniture cream. Br Med J, **i**: 557.

Charde VN, Keskar JB, Ingle AO, Parhad NM, & Daginawala HE (1990) Biodegradation of nitrobenzene by mixed bacterial culture systems. Asian Environ, **12**: 41.

Chiu CW, Lee LH, Wang CY, & Bryan GT (1978) Mutagenicity of some commercially available nitro compounds for *Salmonella typhimurium*. Mutat Res, **58**: 11–22.

Chou WL, Speece RE, & Siddiqi RH (1978) Acclimation and degradation of petrochemical wastewater components by methane fermentation. Biotechnol Bioeng Symp, **8**: 391–414.

Chrastil J & Wilson TJ (1975) 4-Nitrocatechol production from *p*-nitrophenol by rat liver. J Pharmacol Exp Ther, **193**: 631–638.

CIIT (1993) A chronic inhalation toxicity study of nitrobenzene in B6C3F$_1$ mice, Fischer 344 rats and Sprague-Dawley (CD) rats. Vols I & II. Research Triangle Park, North Carolina, Chemical Industry Institute of Toxicology, 20 January 1993 (unpublished).

CLPSD (1988) Contract Laboratory Program Statistical Database. Alexandria, Virginia, Viar and Co., Management Services Division, December 1988.

Cole RH, Frederick RE, Healy RP, & Rolan RG (1984) Preliminary findings of the priority pollutant monitoring project of the nationwide urban runoff program. J Water Pollut Control Fed, **56**: 898–908.

Collins JV, Mayo DR, & Riordan BJ (1982) Economic impact analysis of proposed test rule for nitrobenzene. Washington, DC, US Environmental Protection Agency, Office of Pesticides and Toxic Substances.

Corbett MD & Chipko BR (1977) *N*-Phenylglycolhydroxamate production by action of trans-ketolase on nitrosobenzene. Biochem J, **165**(2): 263–267.

Crebelli R, Conti L, Carere A, Bertoli C, & Del Gicomo N (1995) The effect of fuel composition on the mutagenicity of diesel engine exhaust. Mutat Res, **346**: 167–172.

Crookes MJ, Quarterman P, Howe P, & Dobson S (1994) Environmental hazard assessment: Nitrobenzene. Garston, Watford, United Kingdom Building Research Establishment & Institute of Terrestrial Ecology (Document No. BRE 125/4/32).

Crump KS (1984) A new method for determining allowable daily intakes. Fundam Appl Toxicol, **4**(5): 854–871.

Crump KS (1996) The linearized multistage model and the future of quantitative risk assessment. Hum Exp Toxicol, **15**: 787–798.

Crump KS, Guess HA, & Deal KL (1977) Confidence intervals and test of hypotheses concerning dose response relations inferred from animal carcinogenicity data. Biometrics, **33**: 437–451.

Cunningham ML & Ringrose PS (1983) Benzo(a)pyrene and aniline increase sister chromatid exchanges in cultured rat liver fibroblasts without addition of activating enzymes. Toxicol Lett, **16**: 235–239.

Cupitt LT (1980) Fate of toxic and hazardous materials in the air environment. Washington, DC, US Environmental Protection Agency (Report No. EPA-600/3-80-084; NTIS No. PB80-221948).

Daly J, Jerina D, & Witkop B (1968) Migration of deuterium during hydroxylation of aromatic substrates by liver microsomes. 1. Influence of ring substituents. Arch Biochem Biophys, **128**: 517–527.

Dana MT, Lee RN, & Hales JM (1984) Hazardous air pollutants: Wet removal rates and mechanisms. Research Triangle Park, North Carolina, US Environmental Protection Agency (Report No. EPA-600/3-84-113).

David A, Sroba J, & Hykes P (1965) Biochemische Befunde bei der akuten Vergiftung mit Nitrobenzol bei Menschen. Folia Haemat (Leipzig), **83**: 532–538.

Davis EM, Murray HE, Liehr JG, & Powers EL (1981) Basic microbial degradation rates and chemical byproducts of selected organic compounds. Water Res, **15**: 1125–1127.

Davis EM, Turley JE, Casserly DM, & Guthrie RK (1983) Partitioning of selected organic pollutants in aquatic ecosystems. Biodeterioration, **5**: 176–184.

De Bruin A (1976) Biochemical toxicology of environmental agents. New York, Elsevier/North Holland Inc.

Delclos KB, Tarpley WG, Miller EC, & Miller JA (1984) 4-Aminoazobenzene and *N,N*-dimethyl-4-aminoazobenzene as equipotent hepatic carcinogens in male C57BL x C3H/He F_1 mice and characterization of *N*-(deoxyguanosin-8-yl)-4-aminoazobenzene as the major persistent hepatic DNA-bound dye in these mice. Cancer Res, **44**: 2540–2550.

Dellarco VL & Prival MJ (1989) Mutagenicity of nitro compounds in *Salmonella typhimurium* in the presence of flavin mononucleotide in a preincubation assay. Environ Mol Mutagen, **13**: 116–127.

DeMarini DM, Shelton ML, & Bell DA (1996) Mutation spectra of chemical fractions of a complex mixture: role of nitroarenes in the mutagenic specificity of municipal waste incinerator emissions. Mutat Res, **349**: 1–20.

Deneer JW, Sinnige TL, Seinen W, & Hermens JLM (1987) Quantitative structure–activity relationships for the toxicity and bioconcentration factor of nitrobenzene derivatives towards the guppy (*Poecilia reticulata*). Aquat Toxicol, **10**: 115–129.

Deneer JW, van Leeuwen CJ, Seinen W, Maas-Diepeveen JL, & Hermens JLM (1989) QSAR study of the toxicity and bioconcentration factor of nitrobenzene derivatives towards

Daphnia magna, Chlorella pyrenoidosa and *Photobacterium phosphoreum*. Aquat Toxicol, **15**: 83–98.

Denga N, Moldeus P, Kasilo OM, & Nhachi CFB (1995) Use of urinary *p*-nitrophenol as an index of exposure to parathion. Bull Environ Contam Toxicol, **55**: 296–302.

Dennis RM, Wujeik WJ, Lowe WL, & Marks PJ (1990) Task Order 7. Use of activated carbon for treatment of explosive-contaminated groundwater at the Milan army ammunition plant. Aberdeen Proving Ground, Maryland, US Army Toxic and Hazardous Materials Agency (Report CETHA-TE-CR-90041).

DFG (1995) MAK and BAT values 1995. Deutsche Forschungsgemeinschaft. Weinheim, VCH Verlagsgesellschaft, pp 49, 145 (Report No. 31).

Dietrich DR & Swenberg JA (1991) NCI-Black-Reiter (NBR) male rats fail to develop renal disease following exposure to agents that induce alpha-2u-globulin (alpha 2u) nephropathy. Fundam Appl Toxicol, **16**: 749–762.

Dodd AH (1891) Poisoning by nitro-benzole: Recovery. Br Med J, **i**: 849–850.

Dodd DE, Fowler EH, Snellings WM, Pritts IM, Tyl RW, Lyon JP, O'Neal FO, & Kimmerle G (1987) Reproduction and fertility evaluations in CD rats following nitrobenzene inhalation. Fundam Appl Toxicol, **8**: 493–505.

Dorfman LE & Adams GE (1973) Reactivity of the hydroxyl radical in aqueous solution. Washington, DC, US Department of Commerce, National Bureau of Standards (Report NSRD-NDM-46).

Dorigan J & Hushon J (1976) Air pollution assessment of nitrobenzene. McLean, Virginia, The Mitre Corporation (NTIS No. PB257-776).

Downs TJ, Cifuentes-García E, & Suffet IM (1999) Risk screening for exposure to groundwater pollution in a wastewater irrigation district of the Mexico City region. Environ Health Perspect, **107**: 553–561.

Draper WM & Crosby DG (1984) Solar photooxidation of pesticides in dilute hydrogen peroxide. J Agric Food Chem, **32**: 231–237.

Dreher D & Junod AF (1996) Role of oxygen free radicals in cancer development. Eur J Cancer, **32A**: 30–38.

Dukes MNG ed. (1992) Meyler's side effects of drugs. An encyclopedia of adverse reactions and interactions, 12th ed. London, Elsevier.

Dunkel VC, Pienta R, Sivak A, & Traul KA (1981) Comparative neoplastic transformation responses of Balb/3T3 cells, Syrian hamster embryo cells, and Rauscher murine leukemia virus-infected Fischer 344 rat embryo cells to chemical carcinogens. J Natl Cancer Inst, **67**: 1303–1315.

Dunlap KL (1981) Nitrobenzenes and nitrotoluenes. In: Mark HF, Othmer DF, Overberger CG, Seaborg GT, & Grayson N ed. Kirk-Othmer encyclopedia of chemical technology, vol 15. New York, John Wiley and Sons, pp 916–932.

Dunnivant FM, Schwarzenbach RP, & Macalady DL (1992) Reduction of substituted nitrobenzene in aqueous solutions containing natural organic matter. Environ Sci Technol, **26**: 2133–2141.

ECDIN (2000) Environmental Chemicals Data and Information Network. Environmental Research Program, Joint Research Centre, Commission of European Communities.

Ellenhorn MJ & Barceloux DG (1988) Medical toxicology: Diagnosis and treatment of human poisoning. New York, Elsevier, p 848.

Ellis DD, Jones CM, Larson RA, & Schaeffer DJ (1982) Organic constituents of mutagenic secondary effluents from wastewater treatment plants. Arch Environ Contam Toxicol, **11**: 373–382.

Enfield CG, Walters DM, Wilson JT, & Piwoni MD (1986) Behavior of organic pollutants during rapid-infiltration of wastewater into soil: II. Mathematical description of transport and transformation. Hazard Waste Hazard Mater, **3**: 57–76.

Environment Agency Japan (1992) [Chemicals in the environment. Report on environmental survey and wildlife monitoring of chemicals in fiscal year 1991.] Tokyo, Environment Agency Japan, Department of Environmental Health (in Japanese).

EU (2001) European Union risk assessment report. CAS-No.: 62-53-3, EINECS-No.: 200-539-3. Aniline. Draft. 28 May 2001. Luxembourg, Office for Official Publications of the European Communities.

Ewert R, Buttgereit F, Prugel M, & Reinke P (1998) Intravenous injection of India ink with suicidal intent. Int J Legal Med, **111**: 91–92.

Eyer P (1979) Reactions of nitrosobenzene with reduced glutathione. Chem Biol Interact, **24**: 227–239.

Eyer P & Ascherl M (1987) Reactions of *para*-substituted nitrosoamines with human hemoglobin. Biol Chem Hoppe-Seyler, **368**: 285–294.

Eyer P & Lierheimer E (1980) Biotransformation of nitrobenzene in the red blood cell and the role of glutathione. Xenobiotica, **10**: 517–526.

Eyer P, Kampffmeyer H, Maister H, & Rösch-Oehme E (1980) Biotransformation of nitrosobenzene, phenylhydroxylamine, and aniline in the isolated perfused rat liver. Xenobiotica, **10**: 499–516.

Fairbanks VF & Klee GG (1986) Biochemical aspects of hematology. In: Tietz NW ed. Textbook of clinical chemistry. Philadelphia, Pennsylvania, W.B. Saunders, pp 1536–1540.

Feig DI, Reid TM, & Loeb LA (1994) Reactive oxygen species in tumorigenesis. Cancer Res, **54**: 1890s–1894s.

Feldmann RJ & Maibach HI (1970) Absorption of some organic compounds through the skin in man. J Invest Dermatol, **54**: 399–404.

Fielding M, Gibson TM, James HA, McLoughlin K, & Steel CP (1981) Organic micro-pollutants in drinking water. Medmenham, Water Research Centre (Technical Report TR159).

Fischbach F (1996) A manual of laboratory & diagnostic tests, 5th ed. Philadelphia, Penn-sylvania, Lippincott-Raven Publishers.

Fletcher JS, McFarlane JC, Pfleeger T, & Wickliff C (1990) Influence of root exposure concentration on the fate of nitrobenzene in soybean. Chemosphere, 20: 513–523.

Flohé L, Beckmann R, Giertz H, & Loschen G (1985) Oxygen-centered free radicals as mediators of inflammation. In: Sies H ed. Oxidative stress. New York, Academic Press, pp 403–435.

Flury F & Zernik F (1931) Schädlicke Gase. [Noxious gases.] Berlin, J. Springer [cited in Von Oettingen, 1941].

Fouts JR & Brodie BB (1957) The enzymatic reduction of chloramphenicol, p-nitrobenzoic acid and other nitroaromatic compounds in mammals. J Pharmacol Exp Ther, 119: 197–207.

Freitag D, Geyer H, Kraus A, Viswanathan R, Kotzias D, Attar A, Klein W, & Korte F (1982) Ecotoxicological profile analysis. VII. Screening chemicals for their environmental behaviour by comparative evaluation. Ecotoxicol Environ Saf, 6: 60–81.

Freitag D, Ballhorn L, Geyer H, & Korte F (1985) Environmental hazard profile of organic chemicals. An experimental method for assessment of the behaviour of organic chemicals in the ecosphere by means of simple laboratory tests with ^{14}C-labelled chemicals. Chemo-sphere, 14: 1589–1616.

Fullerton WW (1930) Two rather uncommon fatal cases of poisoning. 1. Nitrobenzene poisoning — suicidal. 2. Sodium fluoride poisoning — accidental. New Engl J Med, 203: 423.

Galloway SM, Armstrong MJ, Reuben C, Colman S, Brown B, Cannon C, Bloom AD, Nakamura F, Ahmed M, Duk S, Rimpo J, Margolin BH, Resnick MA, Anderson B, & Zeiger E (1987) Chromosome aberrations and sister chromatid exchanges in Chinese hamster ovary cells: Evaluations of 108 chemicals. Environ Mol Mutagen, 10 (Suppl 10): 1–175.

Garner RC & Nutman CA (1977) Testing of some azo dyes and their reduction products for mutagenicity using Salmonella typhimurium TA 1538. Mutat Res, 44: 9–19.

Gaylor DW (1989) Preliminary estimates of the virtually safe dose for tumors obtained from the maximum tolerated dose. Regul Toxicol Pharmacol, 9: 101–108.

Gaylor DW (1992) Relationship between the shape of dose–response curves and back-ground tumor rates. Regul Toxicol Pharmacol, 16: 2–9.

Gaylor DW & Gold LS (1995) Quick estimate of the regulatory virtually safe dose based on the maximum tolerated dose for rodent bioassays. Regul Toxicol Pharmacol, 22: 57–63.

Gaylor DW & Gold LS (1998) Regulatory cancer risk assessment based on the quick estimate of a benchmark dose derived from the maximum tolerated dose. Regul Toxicol Pharmacol, 28: 222–225.

Geiger DL, Northcott CE, Call DJ, & Brooke LT (1985) Acute toxicities of organic chemicals to fathead minnows (*Pimephales promelas*). Vol 2. Superior, Wisconsin, University of Wisconsin–Superior, Center for Lake Superior Environmental Studies.

Gershon H, McNeil MW, Parmegiani R, & Godfrey PK (1971) Antifungal activity of substituted nitrobenzenes and anilines. Appl Microbiol, **22**: 438–440.

Gerstle RW (1988) Emissions of trace metals and organic compounds from sewage sludge incineration. In: Proceedings of the 81st Air Pollution Control Association Annual Meeting, Vol 5 (Paper 88/95B.6).

Gerstle R & Carvitti J (1987) Sludge incineration and hazardous air pollutants. Sludge Manage Ser, **17**: 169–174.

Geyer H, Viswanathan R, Freitag D, & Korte F (1981) Relationship between water solubility of organic chemicals and their bioaccumulation by the alga *Chlorella*. Chemosphere, **10**: 1307–1313.

Geyer H, Politzki G, & Freitag D (1984) Prediction of ecotoxicological behaviour of chemicals: relationship between *n*-octanol/water partition coefficient and bioaccumulation of organic chemicals by alga *Chlorella*. Chemosphere, **13**: 269–284.

Gholson AR, Albritton JR, Jayanty RKM, Knoll JE, & Midgett MR (1991) Evaluation of an enclosure method for measuring emissions of volatile organic compounds from quiescent liquid surfaces. Environ Sci Technol, **25**: 519–524.

Gianti SJ, Harkov R, & Bozzelli JW (1984) Monitoring volatile organic compounds at hazardous and sanitary landfills in New Jersey. In: Proceedings of the 77th Air Pollution Control Association Annual Meeting, San Francisco, California, 24–29 June 1984, Vol 1 (Paper 84-3.7).

Gilbert P, Rondelet J, Poncelet F, & Mercier M (1980) Mutagenicity of *p*-nitrosophenol. Food Cosmet Toxicol, **18**: 523–525.

Goldstein A, Aronow L, & Kalman SM (1969) Principles of drug action: The basis of pharmacology. New York, Harper and Row Publishers, pp 274–452.

Goldstein I & Popovici C (1960) Modifications de la catalase et de la péroxidase dans l'intoxication expérimentale au nitrobenzene. Med Lav, **51**: 42–48.

Goldstein RS & Rickert DE (1984) Macromolecular covalent binding of [^{14}C]nitrobenzene in the erythrocyte and spleen of rats and mice. Chem Biol Interact, **50**: 27–37.

Goldstein RS & Rickert DE (1985) Relationship between red blood cell uptake and methemoglobin production by nitrobenzene and dinitrobenzene *in vitro*. Life Sci, **36**: 121–125.

Goldstein RS, Dyroff MC, Rickert DE, & Hamm TE (1983a) Characterization of covalent binding of orally administered nitrobenzene to erythrocytic macromolecules. Pharmacologist, **25**: 116 (Abstr. 78).

Goldstein RS, Irons RD, & Rickert DE (1983b) Pathophysiology of splenic toxicity in mice and rats following oral nitrobenzene (NB) administration. Toxicologist, **3**: 125 (Abstr).

Goldstein RS, Chism JP, & Hamm TJ (1984a) Influence of diet on intestinal microfloral metabolism and toxicity of nitrobenzene. Toxicologist, **4**: 143 (Abstr).

Goldstein RS, Chism JP, Sherrill JM, & Hamm TE (1984b) Influence of dietary pectin on intestinal microfloral metabolism and toxicity of nitrobenzene. Toxicol Appl Pharmacol, **75**: 547–553.

Gomółka B & Gomółka E (1983) Degradation of benzene derivatives in aerated municipal wastewater. Arch Ochr Srodowiska, **3/4**: 113–123.

Gomółka E & Gomółka B (1979) Ability of activated sludge to degrade nitrobenzene in municipal wastewater. Acta Hydrochim Hydrobiol, **7**: 605–622.

Gosselin RE, Smith RP, Hodge HC, & Braddock JE (1984) Clinical toxicology of commercial products, 5th ed. Baltimore, Maryland, Williams & Wilkins, pp II-214, III-31 – III-36.

Govind R, Flaherty PA, & Dobbs RA (1991) Fate and effects of semivolatile organic pollutants during anaerobic digestion of sludge. Water Res, **25**: 547–556.

Grafe E & Homburger A (1914) Industrial nitro-benzene poisoning with Korsakow syndrome, followed by mental debility. Z Gesamte Neurol Psychiat, **25**: 343 [cited in Nabarro, 1948].

Guess H & Crump K (1976) Low-dose extrapolation of data from animal carcinogenesis experiments — analysis of a new statistical technique. Math Biosci, **32**: 15–36.

Guicherit R & Schulting FL (1985) The occurrence of organic chemicals in the atmosphere of the Netherlands. Sci Total Environ, **43**: 193–219.

Gutteridge JMC (1995) Lipid peroxidation and antioxidants as biomarkers for tissue damage. Clin Chem, **41**: 1819–1828.

Guyton KZ & Kensler TW (1993) Oxidative mechanisms in carcinogenesis. Br Med Bull, **49**: 523–544.

Haderlein SB & Schwarzenbach RP (1993) Adsorption of substituted nitrobenzenes and nitrophenols to mineral surfaces. Environ Sci Technol, **27**: 316–326.

Haigler BE & Spain JC (1991) Biotransformation of nitrobenzene by bacteria containing toluene degradative pathways. Appl Environ Microbiol, **57**: 3156–3162.

Hain Z, Dalmacija B, Miskovic D, & Karlovic E (1990) Preparation of drinking water from the surface water of the Danube — A case study. Water Sci Technol, **22**: 253–258.

Hall LW Jr, Hall WS, Bushong SJ, & Herman RL (1987) *In situ* striped bass (*Morone saxatilis*) contaminant and water quality studies in the Potomac River. Aquat Toxicol, **10**: 73–99.

Hallas LE & Alexander M (1983) Microbial transformation of nitroaromatic compounds in sewage effluent. Appl Environ Microbiol, **45**: 1234–1241.

Hamm TE (1984) Ninety-day inhalation study of nitrobenzene in F-344 rats, CD rats and B6C3F₁ mice. Research Triangle Park, North Carolina, Chemical Industry Institute of Toxicology (unpublished).

Hamm TE, Phelps M, Raynor TH, & Irons RD (1984) A 90-day inhalation study of nitrobenzene in F-344 rats, CD rats and B6C3F₁ mice. Toxicologist, **4**: 181 (Abstr).

Hard GC, Rodgers IS, Baetcke KP, Richards WL, McGaughy KP, & Valcovic LP (1993) Hazard evaluation of chemicals that cause accumulation of alpha-2u-globulin, hyaline droplet nephropathy and tubule neoplasia in the kidneys of male rats. Environ Health Perspect, **99**: 313–349.

Hargesheimer EE & Lewis CM (1987) Comparative source water quality monitoring at two surface reservoirs. In: Proceedings of the Annual Conference of the American Water Works Association, Denver, Colorado, pp 153–175.

Harkov R, Kebbekus B, Bozzelli JW, & Lioy PJ (1983) Measurement of selected volatile organic compounds at three locations in New Jersey during the summer season. J Air Pollut Control Assoc, **33**: 1177–1183.

Harkov R, Kebbekus B, Bozzelli JW, Lioy PJ, & Daisey J (1984) Comparison of selected volatile organic compounds during the summer and winter at urban sites in New Jersey. Sci Total Environ, **38**: 259–274.

Harkov R, Gianti SJ Jr, Bozzelli JW, & LaRegina JE (1985) Monitoring volatile organic compounds at hazardous and sanitary landfills in New Jersey. J Environ Sci Health, **A20**: 491–501.

Harmer D, Taylor MJ, Woollen BH, & Loftus MJ (1989) The biological monitoring of workers exposed to nitrobenzene. In: Grandjean EC ed. Occupational health in the chemical industry. XVI Medichem Congress, Helsinki, Finland, 27–30 September 1988. Copenhagen, WHO Regional Office for Europe, pp 142–146.

Harrison MR (1977) Toxic methaemoglobinaemia. A case of acute nitrobenzene and aniline poisoning treated by exchange transfusion. Anaesthesia, **32**: 270–272.

Harton E & Rawl RR (1976) Toxicological and skin corrosion testing of selected hazardous materials. Final report 1973–1974. US Department of Transportation, Office of Hazardous Materials Operations, April (Report No. DOT/MTB/OHMO-76/2; NTIS No. PB-264 975).

Haseman JK & Lockhart A (1994) The relationship between use of the maximum tolerated dose and study sensitivity for detecting rodent carcinogenicity. Fundam Appl Toxicol, **22**(3): 382–391.

Hastings SH & Matsen FA (1948) The photodecomposition of nitrobenzene. J Am Chem Soc, **70**: 3514–3515.

Hattori M, Senoo K, Harada S, Ishizu Y, & Goto M (1984) [The *Daphnia* reproduction test of some environmental chemicals.] Seitai Kagaku, **6**: 23–27 (in Japanese).

Haworth S, Lawlor T, Mortelmans K, Speck W, & Zeiger E (1983) *Salmonella* mutagenicity test results for 250 chemicals. Environ Mutagen, **5** (Suppl 1): 3–38, 46, 114.

Hazleton Laboratories (1982) Final report: 104 week chronic toxicity study in rats. Aniline hydrochloride. Vienna, Virginia, Hazleton Laboratories America Inc. [summary of study reprinted in CIIT Activ, 3(9): 3–4, September 1983].

Hecht SS (1996) Recent studies on mechanisms of bioactivation and detoxification of 4-(methylnitrosamino)-1-(3-pyridyl)-1-butanone (NNK), a tobacco-specific lung carcinogen. CRC Crit Rev Toxicol, **26**: 163–181.

Hecht SS, El-Bayoumy K, Rivenson A, & Fiala ES (1983) Bioassay for carcinogenicity of 3,2'-dimethyl-4-nitrosobiphenyl, o-nitrosotoluene, nitrosobenzene and the corresponding amines in Syrian golden hamsters. Cancer Lett, **20**: 349–354.

Hecht SS, Carmella SG, Foiles PG, & Murphy SE (1994) Biomarkers for human uptake and metabolic activation of tobacco-specific nitrosamines. Cancer Res, **54**: 1912s–1917s.

Hedgecott S & Rogers HR (1991) Freshwater quality. Additional reports undertaken for the Royal Commission on Environmental Pollution. London, HMSO (Report No. CO 2672-M/1).

Heim de Balsac F, Agasse-Lafonte E, & Feil SA (1930) [Professional nitrobenzene poisoning.] Prog Med, **45**: 817–821 (in French).

Hein DW (1988) Acetylator phenotype and arylamine-induced carcinogenesis. Biochem Biophys Acta, **948**: 37–66.

Hein DW, Doll MA, Fretland AJ, Gray K, Deitz AC, Feng Y, Jiang W, Rustan TD, Satran SL, & Wilkie TR Sr (1997) Rodent models of the human acetylation polymorphism: comparisons of recombinant acetyltransferases. Mutat Res, **376**: 101–106.

Heitmuller PT, Hollister TA, & Parrish PR (1981) Acute toxicity of 54 industrial chemicals to sheepshead minnows (*Cyprinodon variegatus*). Bull Environ Contam Toxicol, **27**: 596–604.

Helling CS & Turner BC (1968) Pesticide mobility: Determination by soil thin-layer chromatography. Science, **162**: 562–563.

Helmig D, Muller J, & Klein W (1989) Volatile organic substances in a forest atmosphere. Chemosphere, **19**: 1399–1412.

Hendry DG & Kenley RA (1979) Atmospheric reaction products of organic compounds. Washington, DC, US Environmental Protection Agency (Report No. EPA-560/12-79-001).

Heukelekian H & Rand MC (1955) Biochemical oxygen demand of pure organic compounds. J Water Pollut Control Assoc, **27**: 1040–1053.

Ho CH, Clark BR, & Guerin MR (1981) Analytical and biological analyses of test materials from the synthetic fuel technologies: IV. Studies of chemical structure–mutagenic activity relationships of aromatic nitrogen compounds relevant to synfuels. Mutat Res, **85**: 335–345.

Hogarth CW (1912) A case of poisoning by oil of mirbane (nitro-benzol). Br Med J, i: 183.

Holcombe GW, Phipps GL, Knuth ML, & Felhaber T (1984) The acute toxicity of selected substituted phenols, benzenes and benzoic acid esters to fathead minnows (*Pimephales promelas*). Environ Pollut Ser A, **35**: 367–381.

Holder JW (1998) Nitrobenzene carcinogenicity (CAS No. 98-95-3). Washington, DC, US Environmental Protection Agency, National Center for Environment Assessment (Control No. EPA/600/R-95/100; available for downloading from http://www.epa.gov/ncea or hard copy from the National Technical Information Service, Springfield, Virginia, PB98-143282).

Holder JW (1999a) Nitrobenzene carcinogenicity in animals and human hazard evaluation. Toxicol Ind Health, 15: 445–457.

Holder JW (1999b) Nitrobenzene potential human cancer risk based on animal studies. Toxicol Ind Health, 15: 458–463.

Hoshino M, Akimoto H, & Okuda M (1978) Photochemical oxidation of benzene, toluene and ethylbenzene initiated by OH radicals in the gas phase. Bull Chem Soc Jpn, 51: 718–724.

Howard PH, Boethling RS, Jarvis WF, Meylan WM, & Michalenko EM (1990) Handbook of environmental fate and exposure data for organic chemicals. Chelsea, Michigan, Lewis Publishers.

HSDB (1988) Hazardous substances data bank. Bethesda, Maryland, National Library of Medicine, National Toxicology Information Program.

Huang Q-G, Kong L-R, Liu Y-B, & Wang L-S (1996) Relationship between molecular structure and chromosomal aberrations in *in vitro* human lymphocytes induced by substituted nitrobenzenes. Bull Environ Contam Toxicol, 57: 349–353.

Hughes TJ, Sparacino C, & Frazier S (1984) Validation of chemicals and biological techniques for evaluation of vapors in ambient air/mutagenicity testing of twelve (12) vapor-phase compounds. Research Triangle Park, North Carolina, US Environmental Protection Agency (Report No. EPA 600/1-84-005; NTIS No. PB84-164-219).

Hunt WF Jr, Faoro RB, & Freas W (1986) Report on the interim database for state and local air toxic volatile organic chemical measurements. Washington, DC, US Environmental Protection Agency (Report No. EPA-450/4-86-012).

Hurley R & Testa AC (1966) Photochemical n → π* excitation of nitrobenzene. J Am Chem Soc, 88: 4330–4332.

Hurley R & Testa AC (1967) Nitrobenzene photochemistry 2. Protonation in excited state. J Am Chem Soc, 89(26): 6917–6919.

IARC (1982) Aniline. In: Some aromatic amines, anthraquinones and nitroso compounds, and inorganic fluorides used in drinking water and dental preparations. Lyon, International Agency for Research on Cancer (IARC Monographs on the Evaluation of Carcinogenic Risks to Humans, Vol 27).

IARC (1996) Nitrobenzene. In: Printing processes and printing inks, carbon black and some nitro compounds. Lyon, International Agency for Research on Cancer (IARC Monographs on the Evaluation of Carcinogenic Risks to Humans, Vol 65).

IARC (1999) Species differences in thyroid, kidney and urinary bladder carcinogenesis. Lyon, International Agency for Research on Cancer (IARC Scientific Publications No. 147).

Iida S, Misaka H, & Naya M (1997) A flow cytometric analysis of cytotoxic effects of nitrobenzene on rat spermatogenesis. J Toxicol Sci, **22**: 397–407.

Ikeda M & Kita A (1964) Excretion of *p*-nitrophenol and *p*-aminophenol in the urine of a patient exposed to nitrobenzene. Br J Ind Med, **21**: 210–213.

Ince N (1992) A theoretical approach to determination of evaporative rates and half lives of some priority pollutants in two major lakes of Istanbul. Int J Environ Studies, **40**: 1–12.

IPCS (1994) Assessing human health risks of chemicals: derivation of guidance values for health-based exposure limits. Geneva, World Health Organization, International Programme on Chemical Safety (Environmental Health Criteria 170).

IPCS (2000) Mononitrophenols. Geneva, World Health Organization, International Programme on Chemical Safety (Concise International Chemical Assessment Document No. 20).

Jacobs LW, O'Connor GA, Overcash MA, Zabik MJ, Rygiewicz P, Machno P, Munger S, & Elseewi AA (1987) Effects of trace organics in sewage sludges on soil–plant systems and assessing their risk to humans. In: Land application of sludges: Food chain implications. Chelsea, Michigan, Lewis Publishers.

Jeng CY, Chan DH, & Yaws CL (1992) Data compilation for soil sorption coefficient. Pollut Eng, **24**: 54–60.

Juhnke I & Lüdemann D (1978) Results of research on acute fish toxicity of 200 chemical compounds using golden orfe test. Z Wasser Abwasser Forsch, **11**: 161–164.

Jury WA, Spencer WF, & Farmer WJ (1984) Behavior asessment model for trace organics in soil: III. Application of screening model. J Environ Qual, **13**: 573–579.

Kadlubar FF & Ziegler DM (1974) Properties of a NADH-dependent *N*-hydroxy amine reductase isolated from pig liver microsomes. Arch Biochem Biophys, **162**: 83–92.

Kaiser KLE & Palabrica VS (1991) *Photobacterium phosphoreum* toxicity data index. Water Pollut Res J Can, **26**: 361–431.

Kaiser KLE & Ribo JM (1985) QSAR of toxicity of chlorinated aromatic compounds. In: Tichy M ed. QSAR in toxicology and xenobiochemistry. Amsterdam, Elsevier, pp 27–38.

Kalf GF, Post GB, & Snyder R (1987) Solvent toxicology: Recent advances in the toxicology of benzene, the glycol ethers, and carbon tetrachloride. Annu Rev Pharmacol Toxicol, **27**: 399–427.

Kato M, Kimura H, Hayashi H, Tobe K, Shimizu M, Ota T, & Furuhashi T (1995) Sperm viability in rats treated with nitrobenzene and alpha-chlorohydrin. Teratology, **52**: P-57 (Abstr).

Kawai A, Goto S, Matsumoto Y, & Matsushita H (1987) [Mutagenicity of aliphatic and aromatic nitro compounds.] Jpn J Ind Health, **29**: 34–54 (in Japanese with an English abstract).

Kawashima K, Momma J, Takagi A, Katajima S, & Kurokawa Y (1995a) Examination of sperm motility defects by nitrobenzene with an image processor. Teratology, **52**: P-27 (Abstr).

Kawashima K, Usami M, Sakemi K, & Ohno Y (1995b) Studies on the establishment of appropriate spermatogenic endpoints for male infertility disturbance in rodent induced by drugs and chemicals. I. Nitrobenzene. J Toxicol Sci, **20**: 15–22.

Keher JP (1993) Free radicals as mediators of tissue injury and disease. CRC Crit Rev Toxicol, **23**: 21–48.

Kenley RA, Davenport JE, & Hendry DG (1981) Gas-phase hydroxyl radical reactions. Products and pathways for the reaction of OH with aromatic hydrocarbons. J Phys Chem, **85**: 2740–2746.

Kensler TW, Egner PA, Taffe BG, & Trush MA (1989) Role of free radicals in tumor promotion and progression. Prog Clin Biol Res, **298**: 233–248.

Khan MF, Boor PJ, Gu Y, Alcock NW, & Ansari GA (1997) Oxidative stress in the spleno-toxicity of aniline. Fundam Appl Toxicol, **35**: 22–30.

Kiese M (1966) The biochemical production of ferrihemoglobin forming derivatives from aromatic amines and mechanisms of ferrihemoglobin formation. Pharmacol Rev, **18**: 1091–1161.

Kiese M (1974) Methemoglobinemia: A comprehensive treatise. Cleveland, Ohio, CRC Press.

Kincannon DF & Lin YS (1985) Microbial degradation of hazardous wastes by land treatment. In: Bell JM ed. Proceedings of the 40th Industrial Waste Conference. Boston, Massachusetts, Butterworth-Heinemann, pp 607–619.

King CM, Romano JL, & Schuetzle D ed. (1988) Carcinogenic and mutagenic responses to aromatic amines and nitroamines. New York, Elsevier Publishing Co.

Klaassen CD ed. (2001) Casarett & Doull's toxicology: the basic science of poisons, 6th ed. New York, McGraw-Hill.

Kligerman AD, Erexson GL, Wilmer JL, & Phelps MC (1983) Analysis of cytogenetic damage in rat lymphocytes following *in vivo* exposure to nitrobenzene. Toxicol Lett, **18**: 219–226.

Koch R & Nagel M (1988) Quantitative structure activity relationships in soil ecotoxicology. Sci Total Environ, **77**: 269–276.

Kodama M, Kaneko M, Aida M, Inoue F, Nakayama T, & Akimoto H (1997) Free radical chemistry of cigarette smoke and its implication in human cancer. Anticancer Res, **17**: 433–437.

Kopfler FC, Melton RG, Mullaney JL, & Tardiff RG (1977) Human exposure to water pollutants. Adv Environ Sci Technol, **8**: 419–433.

Korte F & Klein W (1982) Degradation of benzene in the environment. Ecotoxicol Environ Saf, 6: 311–327.

Krewski D, Gaylor DW, Soms AP, & Szyszkowicz M (1993) An overview of the report — correlation between carcinogenic potency and the maximum tolerated dose — implications for risk assessment. Risk Anal, 13(4): 383–398.

Kubota Y (1979) Experience with the chemical substances control law in Japan. Ecotoxicol Environ Saf, 3: 256–268.

Kühn R, Pattard M, Pernak KD, & Winter A (1989) Results of the harmful effects of water pollutants to *Daphnia magna* in the 21 day reproduction test. Water Res, 23: 501–510.

Kuhn W & Clifford D (1986) Experience with specific organic analyses for water quality control in West Germany. In: Proceedings of the 13th American Water Works Association Water Quality Technology Conference. Denver, Colorado, American Water Works Association, pp 353–363.

Kumar A, Chawla R, Ahuja S, Girdhar KK, & Bhattacharya A (1990) Nitrobenzene poisoning and spurious pulse oximetry. Anaesthesia, 45: 949–951.

Kuroda Y (1986) Genetic and chemical factors affecting chemical mutagenesis in cultured mammalian cells. In: Shankel DM, Hartman PE, Kada T, Hollaender A, Wilson CM, & Kuny G ed. Antimutagenesis and anticarcinogenesis mechanisms. New York, Plenum Press, pp 359–375.

Laramée JA, Kocher CA, & Deinzer ML (1992) Application of a trochoidal electron monochromator/mass spectrometer system to the study of environmental chemicals. Anal Chem, 64: 2316–2322.

LaRegina J, Bozzelli JW, Harkov R, & Gianti S (1986) Volatile organic compounds at hazardous waste sites and a sanitary landfill in New Jersey. Environ Prog, 5: 18–27.

Lauwerys RR (1991) Occupational toxicology. In: Amdur MO, Doull J, & Klaassen CD ed. Casarett & Doull's toxicology — The basic science of poisons, 4th ed. New York, Pergamon Press, pp 947–969.

Leader SD (1932) Nitrobenzene poisoning: Report of an unusual case in a child. Arch Pediatr, 49: 245–250.

LeBlanc GA (1980) Acute toxicity of priority pollutants to water flea (*Daphnia magna*). Bull Environ Contam Toxicol, 24: 684–691.

Levin AA & Dent JG (1982) Comparison of the metabolism of nitrobenzene by hepatic microsomes and cecal microflora from Fischer-344 rats *in vitro* and the relative importance of each *in vivo*. Drug Metab Dispos, 10: 450–454.

Levin AA, Bus JS, & Dent JG (1982) Interactions of nitrobenzene with hepatic microsomes. Evidence for cytochrome P-450 uncoupling. Toxicologist, 2: 17.

Levin AA, Bosakowski T, Earle LL, & Butterworth BE (1988) The reversibility of nitrobenzene-induced testicular toxicity: continuous monitoring of sperm output from vasocystotomized rats. Toxicology, 53: 219–230.

Levins P, Adams J, Brenner P, Coons S, & Harris G (1979) Sources of toxic pollutants found in influents to sewerage treatment plants. Boston, Massachusetts, AD Little Inc. (ADL 81099-63).

Lewalter J & Ellrich D (1991) Nitroaromatic compounds (nitrobenzene, *p*-nitrotoluene, *p*-nitrochlorobenzene, 2,6-dinitrotoluene, *o*-dinitrobenzene, 1-nitronaphthalene, 2-nitronaphthalene, 4-nitrobiphenyl). Anal Hazard Subst Biol Mater, 3: 207–215.

Liepins R, Mixon F, Hudak C, & Parsons TB (1977) Industrial process profiles for environmental use. Chapter 6. The industrial organic chemicals industry. Washington, DC, US Environmental Protection Agency (Report No. EPA-600/2-77/023F).

Linch AL (1974) Biological monitoring for industrial exposure to cyanogenic aromatic nitro and amino compounds. Am Ind Hyg Assoc J, 35: 426–432.

Linder RE, Strader LF, Slott VL, & Suarez JD (1992) Endpoints of spermatotoxicity in the rat after short duration exposures to fourteen reproductive toxicants. Reprod Toxicol, 6: 491–505.

Lioy PJ, Daisey JM, Atherholt T, Bozzelli J, Darack R, Fisher R, Greenberg A, Harkov R, Kebbekus B, Kneip TJ, Lewis J, McGarrity G, McGeorge L, & Reiss NM (1983) The New Jersey Project on airborne toxic elements and organic substances (ATEOS): A summary of the 1981 summer and 1982 winter studies. J Air Pollut Control Fed, 33: 649–657.

Locket S (1957) Clinical toxicology. London, Henry Kimpton, pp 505–506.

Løkke H (1984) Sorption of selected organic pollutants in Danish soils. Ecotoxicol Environ Saf, 8: 395–409.

Lu PY & Metcalf RL (1975) Environmental fate and biodegradability of benzene derivatives as studied in a model aquatic ecosystem. Environ Health Perspect, 10: 269–284.

Luke JSH & Betton GR (1987) *In vitro* evaluation of haemic systems in toxicology. In: Atterwill CK & Steele CE ed. *In vitro* methods in toxicology. Cambridge, Cambridge University Press.

Lutin PA, Cibulka JJ, & Malaney GW (1965) Oxidation of selected carcinogenic compounds by activated sludge. Purdue Univ Eng Ext Ser, 118: 131–145.

Lyman WJ, Reehl WF, & Rosenblatt DH (1982) Handbook of chemical property estimation methods. New York, McGraw-Hill Book Co., p 960.

Maas-Diepeveen JL & Van Leeuwen CJ (1986) [Aquatic toxicity of aromatic nitro compounds and anilines to several freshwater species.] Arnhem, Institute for Inland Water Management and Waste Water Treatment, Laboratory for Ecotoxicology, 10 pp (Report No. 86-42) (in Dutch).

Mabey WR, Smith JH, Podoll RT, Johnson HL, Mill T, Chou T-W, Gates J, Waight Partridge I, Jaber H, & Vandenberg D (1982) Aquatic fate process data for organic priority pollutants. Washington, DC, US Environmental Protection Agency, Office of Water Regulations and Standards (Report No. EPA 440/4-81-014; NTIS No. PB87-169090/XAB).

Mallevialle J, Bruchet A, & Schmitt E (1984) Nitrogenous organic compounds: Identification and significance in several French water treatment plants. In: Proceedings of the 11th American Water Works Association Water Quality Technology Conference, pp 83–96.

Manufacturing Chemists Association (1968) Research on chemical odors. Part 1. Odor thresholds for 53 commercial chemicals. Washington, DC, Manufacturing Chemists Association, October 1968, p 23.

Maples KR, Eyer P, & Mason RP (1990) Aniline-, phenylhydroxylamine-, nitrosobenzene-, and nitrobenzene-induced hemoglobin thiyl free radical formation *in vivo* and *in vitro*. Mol Pharmacol, **37**: 311–318.

Marchini S, Tosato ML, Norberg-King TJ, Hammermeister DE, & Hoglund MD (1992) Lethal and sublethal toxicity of benzene derivatives to the fathead minnow, using a short term test. Environ Toxicol Chem, **11**: 187–195.

Marchini S, Hoglund MD, Broderius SJ, & Tosato ML (1993) Comparison of the susceptibility of daphnids and fish to benzene derivatives. In: Proceedings of the 2nd European Conference on Ecotoxicology, Recent Advances in Toxicology, 11–15 May 1992, pp. 799–808. Sci Total Environ Suppl 1993, Part 1.

Marion CV & Malaney GW (1963) Ability of activated sludge microorganisms to oxidize aromatic organic compounds. Purdue Univ Eng Ext Ser, **115**: 297–308.

Maser E (1997) Stress, hormonal changes, alcohol, food constituents and drugs: factors that advance the incidence of tobacco smoke-related cancer? Trends Pharmacol Sci, **18**: 270–275.

Mason RP (1982) Free-radical intermediates in the metabolism of toxic chemicals. In: Pryor WA ed. Free radicals in biology, Vol V. New York, Academic Press, pp 161–222.

Mason RP & Holtzman JL (1975a) The mechanism of microsomal and mitochondrial nitroreductase. Electron spin resonance evidence for nitroaromatic free radical intermediates. Biochemistry, **14**: 1626–1632.

Mason RP & Holtzman JL (1975b) The role of catalytic superoxide formation in the O_2 inhibition of nitroreductase. Biochem Biophys Res Commun, **67**: 1267–1274.

Matsumaru H & Yoshida T (1959) Experimental studies of nitrobenzol poisoning. Kyushu J Med Sci, **10**: 259–264.

McCann J, Choi E, Yamasaki E, & Ames BN (1975) Detection of carcinogens as mutagens in the *Salmonella*/microsome test; assay of 300 chemicals. Proc Natl Acad Sci USA, **12**: 5135.

McCrady JK, McFarlane C, & Lindstrom FT (1987) The transport and affinity of substituted benzenes in soybean stems. J Exp Bot, **38**: 1875–1890.

McFarlane C, Nolt C, Wickliff C, Pfleeger T, Shimabuku R, & McDowell M (1987a) The uptake, distribution and metabolism of four organic chemicals by soybean plants and barley roots. Environ Toxicol Chem, **6**: 847–856.

McFarlane JC, Pfleeger T, & Fletcher J (1987b) Transpiration effect on the uptake and distribution of bromacil, nitrobenzene and phenol in soybean plants. J Environ Qual, **16**: 372–376.

McLaren TT, Foster PM, & Sharpe RM (1993a) Effect of age on seminiferous tubule protein secretion and the adverse effects of testicular toxicants in the rat. Int J Androl, **16**: 370–379.

McLaren TT, Foster PM, & Sharpe RM (1993b) Identification of stage-specific changes in protein secretion by isolated seminiferous tubules from the rat following exposure to either *m*-dinitrobenzene or nitrobenzene. Fundam Appl Toxicol, **21**: 384–392.

Medinsky MA & Irons RD (1985) Sex, strain and species differences in the response of rodents to nitrobenzene vapours. Toxicity of nitroaromatic compounds. New York, Hemisphere Publishing Corporation.

Meijers AP & Van Der Leer RC (1976) The occurrence of organic micropollutants in the River Rhine and the River Maas in 1974. Water Res, **10**: 597–604.

Miller JA (1970) Carcinogenesis by chemicals: an overview — G.H.A. Clowes memorial lecture. Cancer Res, **30**: 559–576.

Mirsalis JC, Tyson CK, & Butterworth BE (1982) Detection of genotoxic carcinogens in the *in vivo–in vitro* hepatocyte DNA repair essay. Environ Mutagen, **4**: 553–562.

MITI (1992) Biodegradation and bioaccumulation data of existing chemicals based on the CSCL Japan. Tokyo, Ministry of International Trade & Industry, Japan Chemicals Inspection & Testing Institute, Japan Chemical Industry Ecology–Toxicology & Information Center.

Mitsumori K, Kodama Y, Uchida O, Takada K, Saito M, Naito K, Tanaka S, Kurokawa Y, Usami M, Kawashima K, Yasuhara K, Toyoda K, Onodera H, Furukawa F, Takahashi M, & Hayashi Y (1994) Confirmation study, using nitrobenzene, of the combined repeat dose and reproductive/developmental toxicity test protocol proposed by the Organisation for Economic Cooperation and Development (OECD). J Toxicol Sci, **19**: 141–149.

Moeschlin S (1965) Poisoning: Diagnosis and treatment, 1st US ed. New York, Grune & Stratton, pp 362–366.

Morgan KT, Gross EA, Lyght O, & Bond JA (1985) Morphologic and biochemical studies of a nitrobenzene-induced encephalopathy in rats. Neurotoxicology, **6**: 105–116.

Morrissey RE, Schwetz BA, Lamb JC, Ross MD, Teague JL, & Morris RW (1988) Evaluation of rodent sperm, vaginal cytology, and reproductive organ weight data from National Toxicology Program 13 week studies. Fundam Appl Toxicol, **11**: 343–358.

Muehlberger CW (1925) Shoe dye poisoning. J Am Med Assoc, **84**: 1987–1991.

Myślak Z, Piotrowski JK, & Musialowicz E (1971) Acute nitrobenzene poisoning: A case report with data on urinary excretion of *p*-nitrophenol and *p*-aminophenol. Arch Toxikol, **28**: 208–213.

Nabarro JD (1948) A case of acute mononitrobenzene poisoning. Br Med J, **15**: 929–931.

Narayanan B, Suidan MT, Gelderloos AB, & Brenner RC (1993) Treatment of semi-volatile compounds in high strength wastes using an anaerobic expanded-bed GAC reactor. Water Res, **27**: 171–180.

Nelson CR & Hites RA (1980) Aromatic amines in and near the Buffalo River. Environ Sci Technol, **14**: 1147–1149.

Netke SP, Roomi MW, Tsao C, & Niedzwiecki A (1997) Ascorbic acid protects guinea pigs from acute aflatoxin toxicity. Toxicol Appl Pharmacol, **143**: 429–435.

Neuhauser EF, Loehr RC, Malecki MR, Milligan DL, & Durkin PR (1985) The toxicity of selected organic chemicals to the earthworm *Eisenia fetida*. J Environ Qual, **14**: 383–388.

Neuhauser EF, Loehr RC, & Malecki MR (1986) Contact and artificial soil tests using earthworms to evaluate the impact of wastes in soil. In: Hazardous and industrial solid waste testing: Fourth symposium. Philadelphia, Pennsylvania, American Society for Testing and Materials, pp 192–203 (ASTM STP 886).

Neumann H-G (1984) Analysis of hemoglobin as a dose monitor for alkylating and arylating agents. Arch Toxicol, **56**: 1–6.

Neumann H-G (1988) Biomonitoring of aromatic amines and alkylating agents by measuring haemoglobin adducts. Int Arch Occup Environ Health, **60**: 151–155.

Nielsen IR, Diment J, & Dobson S (1993) Environmental hazard assessment: Aniline. Watford, Building Research Establishment (TSD/14).

NIOSH (1977) Nitrobenzene — method S217. In: NIOSH manual of analytical methods, Vol 3. Cincinnati, Ohio, National Institute of Occupational Safety and Health, pp S217-1–S217-9.

NIOSH (1984) Nitrobenzene — method 2005. In: NIOSH manual of analytical methods, 3rd ed. Cincinnati, Ohio, National Institute of Occupational Safety and Health, pp 2005-1–2005-5.

NIOSH (1990) National Occupational Exposure Survey (1981–1983). Cincinnati, Ohio, National Institute of Occupational Safety and Health (unpublished provisional data as of 7 January 1990).

Nishino SF & Spain JC (1993) Degradation of nitrobenzene by a *Pseudomonas pseudoalcaligenes*. Appl Environ Microbiol, **59**: 2520–2525.

Nohmi T, Yoshikawa K, Nakadate M, Miyata R, & Ishidate M Jr (1984) Mutations in *Salmonella typhimurium* and inactivation of *Bacillus subtilis* transforming DNA induced by phenylhydroxylamine derivatives. Mutat Res, **136**: 159–168.

Nojima K & Kanno S (1977) Studies on photochemistry of aromatic hydrocarbons. IV. Mechanism of formation of nitrophenols by the photochemical reaction of benzene and toluene with nitrogen oxides in air. Chemosphere, 6: 371–376.

Nolt CL (1988) Uptake and translocation of six organic chemicals in a newly-designed plant exposure system and evaluation of plant uptake aspects of the prebiologic screen for ecotoxicologic effects. MS Thesis. Ithaca, New York, Cornell University.

Noordsij A, Puyker LM, & Van Der Gaag MA (1985) The quality of drinking water prepared from bank-filtered river water in the Netherlands. Sci Total Environ, 47: 273–292.

Norambuena E, Videla LA, & Lissi EA (1994) Interaction of nitrobenzoates with haemoglobin in red blood cells and a haemolysate. Human Exp Toxicol, 13: 345–351.

NTP (1983a) Nitrobenzene (14-day and 90-day gavage studies) in Fischer 344 rats and B6C3F$_1$ mice. Conducting Laboratory: EG & G Mason Research Institute. Reviewed by the Pathology Working Group, National Toxicology Program, Public Health Service, National Institutes of Health, 20 July 1983 (unpublished).

NTP (1983b) Nitrobenzene (14-day and 90-day skin paint studies) in Fischer 344 rats and B6C3F$_1$ mice. Conducting Laboratory: EG & G Mason Research Institute. Reviewed by the Pathology Working Group, National Toxicology Program, Public Health Service, National Institutes of Health, 15 July 1983 (unpublished).

OECD (1981) OECD guidelines for testing of chemicals. Paris, Organisation for Economic Co-operation and Development.

OECD (1992) Report of the OECD workshop on extrapolation of laboratory aquatic toxicity data to the real environment. Paris, Organisation for Economic Co-operation and Development (OECD Environment Monograph No. 59).

OECD (1995) Guidance document for aquatic effects assessment. Paris, Organisation for Economic Co-operation and Development (OECD Environment Monograph No. 92).

Ohkuma Y & Kawanishi S (1999) Oxidative DNA damage by a metabolite of carcinogenic and reproductive toxic nitrobenzene in the presence of NADH and Cu(II). Biochem Biophys Res Commun, 257: 555–560.

Ongley ED, Birkholz DA, Carey JH, & Samoiloff MR (1988) Is water a relevant sampling medium for toxic chemicals? An alternative environmental sensing strategy. J Environ Qual, 17: 391–401.

Otson R, Williams DT, & Bothwell PD (1982) Volatile organic compounds in water at thirty Canadian potable water treatment facilities. J Assoc Off Anal Chem, 65: 1370–1374.

Pacséri I, Magos L, & Bátskor IA (1958) Threshold and toxic limits of some amino and nitro compounds. Am Med Assoc Arch Ind Health, 18: 1–8.

Parke DV (1956) Studies in detoxication. 68. The metabolism of [^{14}C]nitrobenzene in the rabbit and guinea pig. Biochem J, 62: 339–346.

Parkes WE & Neill DW (1953) Acute nitrobenzene poisoning with transient amino-aciduria. Br Med J, i: 653–655.

Parodi S, Taningher M, Russo P, Pala M, Tamaro M, & Monti-Bragadin C (1981) DNA-damaging activity *in vivo* and bacterial mutagenicity of sixteen aromatic amines and azo-derivatives, as related quantitatively to their carcinogenicity. Carcinogenesis, 2: 1317–1326.

Parodi S, Pala M, Russo P, Zunino A, Balbi C, Albini A, Valerio F, Cimberle MR, & Santi L (1982) DNA damage in liver, kidney, bone marrow, and spleen of rats and mice treated with commercial and purified aniline as determined by alkaline elution assay and sister chromatid exchange induction. Cancer Res, 42: 2277–2283.

Patil SF & Lonkar ST (1992) Thermal desorption–gas chromatography for the determination of benzene, aniline, nitrobenzene and chlorobenzene in workplace air. J Chromatogr, 600: 344–351 [Published erratum appears in J Chromatogr, 625(2): 402, 1992].

Patil SS & Shinde VM (1988) Biodegradation studies of aniline and nitrobenzene in aniline plant waste water by gas chromatography. Environ Sci Technol, 22: 1160–1165.

Patil SS & Shinde VM (1989) Gas chromatographic studies on the biodegradation of nitrobenzene and 2,4-dinitrophenol in the nitrobenzene plant wastewater. Environ Pollut, 57: 235–250.

Pellizzari ED (1978) Quantification of chlorinated hydrocarbons in previously collected air samples. Research Triangle Park, North Carolina, US Environmental Protection Agency (Report No. EPA-450/3-78-112).

Pendergrass SM (1994) An approach to estimating workplace exposure to o-toluidine, aniline and nitrobenzene. Am Ind Hyg Assoc J, 55: 733–737.

Perry DL, Chuang CC, Jungclaus GA, & Warner JS (1979) Identification of organic compounds in industrial effluent discharges. Athens, Georgia, US Environmental Protection Agency, Office of Research and Development (Report No. EPA-600/4-79-016; NTIS No. PB-294 794/3).

Peterson PJ, Mason RP, Hovsepian J, & Holtzman JL (1979) Oxygen-sensitive and -insensitive nitroreduction by *Escherichia coli* and rat hepatic microsomes. J Biol Chem, 254: 4009–4014.

Piet GJ, Morra CH, & De Kruijf HA (1981) The behavior of organic micropollutants during passage through the soil. Stud Environ Sci, 17: 557–564.

Piotrowski J (1967) Further investigations on the evaluation of exposure to nitrobenzene. Br J Ind Med, 24: 60–65.

Pitter P (1976) Determination of biological degradability of organic substances. Water Res, 10: 231–235.

Piwoni MD, Wilson JT, Walters DM, Wilson BH, & Enfield CG (1986) Behaviour of organic pollutants during rapid-infiltration of wastewater into soil: I. Processes, definition, and characterization using a microcosm. Hazard Waste Hazard Mater, 3: 43–55.

Pizzolatti MG & Yunes RA (1990) Azoxybenzene formation from nitrosobenzene and phenyl-hydroxylamine — a unified view of the catalysis and mechanisms of the reactions. J Chem Soc Perkin Trans, 2(5): 759–764.

Polson CJ & Tattersall RN (1969) Clinical toxicology, 2nd ed. Bath, Pitman Press.

Pope AA, Cruse PA, & Most CC (1988) Toxic air pollutant emission factors — A compilation for selected air toxic compounds and sources. Washington, DC, US Environmental Protection Agency (Report No. EPA-450/2-88-006A).

Price CJ, Tyl RW, Marks TA, Paschke LL, Ledoux TA, & Reel JR (1985) Teratologic and postnatal evaluation of aniline hydrochloride in the Fischer 344 rat. Toxicol Appl Pharmacol, **77**: 465–478.

Probst GS, McMahon RE, Hill LE, Thompson CZ, Epp JK, & Neal SB (1981) Chemically-induced unscheduled DNA synthesis in primary rat hepatocyte cultures: A comparison with bacterial mutagenicity using 218 compounds. Environ Mutagen, **3**: 11–32.

Pryor WA (1997) Cigarette smoke radicals and the role of free radicals in chemical carcino-genicity. Environ Health Perspect, **105**: 875–882.

Purchase IFH, Longstaff E, Ashby J, Styles JA, Anderson D, Lefevre PA, & Westwood FR (1978) An evaluation of 6 short term tests for detecting organic chemical carcinogens. Br J Cancer, **37**: 873–903.

Ramos EU, Vaes WHJ, Verhaar HJM, & Hermens JLM (1998) Quantitative structure–activity relationships for the aquatic toxicity of polar and nonpolar narcotic pollutants. J Chem Inf Comp Sci, **38**(5): 845–852.

Ramos EU, Vaes WHJ, Mayer P, & Hermens JLM (1999) Algal growth inhibition of *Chlorella pyrenoidosa* by polar narcotic pollutants: toxic cell concentrations and QSAR modeling. Aquat Toxicol, **46**(1): 1–10.

Ramsay DHE & Harvey CC (1959) Marking ink poisoning. An outbreak of methemo-globinemia cyanosis in newborn babies. Lancet, **i**: 910–912.

Rapoport IA (1965) Mutational effect of paranitroaceto-phenylene-tri-phenylphosphine in connection with the additivity of mutational tendencies. Presented by Academician N. N. Semenove, 21 August 1964. Translated from Doklady Akademii Nauk SSR, **60**: 707–709.

Reddy BG, Pohl LR, & Krishna G (1976) The requirement of the gut flora in nitrobenzene-induced methemoglobinemia in rats. Biochem Pharmacol, **25**: 1119–1122.

Rejsek K (1947) *m*-Dinitrobenzene poisoning. Mobilisation by alcohol and sunlight. Acta Med Scand, **127**: 179–191.

Rickert DE (1984) Toxicity of nitroaromatic compounds. Washington, DC, Hemisphere Publishing Corp.

Rickert DE (1987) Metabolism of nitroaromatic compounds. Drug Metab Rev, **18**: 23–53.

References

Rickert DE, Bond JA, Long RM, & Chism JP (1983) Metabolism and excretion of nitrobenzene by rats and mice. Toxicol Appl Pharmacol, 67: 206–214.

Robinson D, Smith JN, & Williams RT (1951) Studies in detoxication 40. The metabolism of nitrobenzene in the rabbit. o-, m- and p-nitrophenols, o-, m- and p-aminophenols and 4-nitrocatechol as metabolites of nitrobenzene. Biochem J, 50: 228–238.

Rogozen MB, Rich HE, Guttman MA, & Grosjean D (1987) Evaluation of potential toxic air contaminants. Phase 1. Final report. Sacramento, California, State of California Air Resources Board (Report ARB/R-88/333).

Romero IA, Ray DE, Chan MWK, & Abbott NJ (1996) An in vitro study of m-dinitrobenzene toxicity on the cellular components of the blood–brain barrier, astrocytes and endothelial cells. Toxicol Appl Pharmacol, 139: 94–101.

Rosen ME, Pankow JF, Gibs J, & Imbrigiotta TE (1992) Comparison of downhole and surface sampling for the determination of volatile organic compounds (VOCs) in ground water. Ground Water Monit Rev, 12: 126–133.

Rosenkranz HS (1996) Mutagenic nitroarenes, diesel emissions, particulate-induced mutations and cancer: an assay on cancer-causation by a moving target. Mutat Res, 367: 65–72.

Rosenkranz HS & Mermelstein R (1983) Mutagenicity and genotoxicity of nitroarenes. All nitro-containing chemicals were not created equal. Mutat Res, 114: 217–267.

Roy WR & Griffin RA (1985) Mobility of organic solvents in water-saturated soil materials. Environ Geol Water Sci, 7: 241–247.

Sabbioni G (1994) Hemoglobin binding of nitroarenes and quantitative structure–activity relationships. Chem Res Toxicol, 7: 267–274.

Salmowa J, Piotrowski J, & Neuhorn U (1963) Evaluation of exposure to nitrobenzene: Absorption of nitrobenzene vapour through lungs and excretion of p-nitrophenol in urine. Br J Ind Med, 20: 41–46.

Sanders FG (1920) Nitrobenzene poisoning with cyanosis: Report of case. J Am Med Assoc, 74: 1518–1519.

Schimelman MA, Soler JM, & Muller HA (1978) Methemoglobinemia: nitrobenzene ingestion. JACEP, J Am Coll Emerg Physicians, 7: 406–408.

Schmieder PK & Henry TR (1988) Plasma binding of 1-butanol, phenol, nitrobenzene and pentachlorophenol in the rainbow trout and rat: a comparative study. Comp Biochem Physiol, Part C, 91: 413–418.

Schroeder RE, Terrill JB, Lyon JP, Kaplan AM, & Kimmerle G (1986) An inhalation teratology study in the rabbit with nitrobenzene. Toxicologist, 6: 93 (Abstr).

Schultz TW, Dawson DA, & Lin DT (1989) Comparative toxicity of selected nitrogen-containing aromatic compounds in the Tetrahymena pyriformis and Pimephales promelas test systems. Chemosphere, 18: 2283–2291.

231

Schwarzenbach RP, Stierli R, Lanz K, & Zeyer J (1990) Quinone and iron porphyrin mediated reduction of nitroaromatic compounds in homogeneous aqueous solution. Environ Sci Technol, **24**: 1566–1574.

Scott RW & Hanzlik PJ (1920) Poisoning by alcohol "denatured" with nitrobenzene. J Am Med Assoc, **10**: 1000.

Sealy RC, Swartz HM, & Olive PL (1978) Electron spin resonance–spin trapping. Detection of superoxide formation during aerobic microsomal reduction of nitro-compounds. Biochem Biophys Res Commun, **82**: 680–684.

Seip HM, Alstad J, Carlberg GE, Martinsen K, & Skaane R (1986) Measurement of mobility of organic compounds in soils. Sci Total Environ, **50**: 87–101.

Shackelford WM, Cline DM, Faas L, & Kurth G (1983) An evaluation of automated spectrum matching for survey identification of wastewater components by gas chromatography–mass spectrometry. Anal Chim Acta, **146**: 15–27.

Shimizu M & Yano E (1986) Mutagenicity of mono-nitrobenzene derivatives in the Ames test and rec assay. Mutat Res, **170**: 11–22.

Shimkin MB (1939) Acute toxicity of mononitrobenzene in mice. Proc Soc Exp Biol Med, **42**: 844–846.

Shimo T, Onodera H, Matsushima Y, Todate A, Mitsumori K, Maekawa A, & Takahashi M (1994) [A 28-day repeated dose toxicity study of nitrobenzene in F344 rats.] Eisei Shikenjo Hokoku, **112**: 71–81 (in Japanese with an English abstract).

Shinoda K, Mitsumori K, Yasuhara K, Ueyama C, Onodera H, Takegawa K, Takahashi M, & Umemura T (1998) Involvement of apoptosis in the rat germ cell degeneration induced by nitrobenzene. Arch Toxicol, **72**: 296–302.

Simmons MS & Zepp RG (1986) Influence of humic substances on photolysis of nitro-aromatic compounds in aqueous systems. Water Res, **20**: 899–904.

Simpson K ed. (1965) Nitrobenzene, oil of mirbane ($C_6H_5NO_2$), and dinitrobenzene ($C_6H_4(NO_2)_2$). Taylor's principles and practice of medical jurisprudence, 12th ed. London, JA Churchill, vol 2, pp 441–443.

Smith RP, Alkaitis AA, & Shafer PR (1967) Chemically induced methemoglobinemias in the mouse. Biochem Pharmacol, **16**: 317–328.

Smith RV & Rosazza JP (1974) Microbial models of mammalian metabolism. Aromatic hydroxylation. Arch Biochem Biophys, **161**: 551–558.

Smith S & Fiddes FS (1955) Forensic medicine. London, Churchill.

Smyth HF, Weil CS, West JS, & Carpenter CP (1969) An exploration of joint toxic action: Twenty seven industrial chemicals intubated in rats in all possible pairs. Toxicol Appl Pharmacol, **14**: 340–347.

Spicer CW, Riggin RM, Holdren MW, DeRoos FL, & Lee RN (1985) Atmospheric reaction products from hazardous air pollutant degradation. Washington, DC, US Environmental Protection Agency (Report No. EPA-600/3-85-028).

Spielmann H, Gerner I, Kalweit S, Moog R, Wirnsberger T, Krauser K, Kreiling R, Kreuzer H, Lüpke N-P, Miltenburger HG, Müller N, Mürmann P, Pape W, Siegemund B, Spengler J, Steiling W, & Wiebel FJ (1991) Interlaboratory assessment of alternatives to the Draize eye irritation test in Germany. Toxic In Vitro, 5: 539–542.

Spinner JR (1917) Nitrobenzol not an abortifacient. J Am Med Assoc, 69: 2155.

SRI (1985) Directory of chemical producers. Western Europe. Menlo Park, California, SRI International.

Staples CA, Werner AF, & Hoogheem TJ (1985) Assessment of priority pollutant concentrations in the United States using STORET database. Environ Toxicol Chem, 4: 131–142.

Staretz ME, Murphy SE, Patten CJ, & Nunes MG (1997) Comparative metabolism of the tobacco-related carcinogens benzo[a]pyrene, 4-(methylnitrosamino)-1-(3-pyridyl)-1-butanone, 4-(methylnitrosamino)-1-(3-pyridyl)-1-butanol and N'-nitrosonornicotine in human hepatic microsomes. Drug Metab Dispos, 25: 154–162.

Stevens AM (1928) Cyanosis in infants from nitrobenzene. J Am Med Assoc, 90: 116.

Stier A, Clauss R, & Lücke A (1980) Redox cycle of stable mixed nitroxides formed from carcinogenic aromatic amines. Xenobiotica, 10: 661–673.

Stifel RE (1919) Methemoglobinemia due to poisoning by shoe dye: Report of a series of cases at an army camp. J Am Med Assoc, 72: 395–396.

Stolk JM & Smith RP (1966) Species differences in methemoglobin reductase activity. Biochem Pharmacol, 15: 343–351.

Stover EL & Kincannon DF (1983) Biological treatability of specific organic compounds found in chemical industry wastewaters. J Water Pollut Control Fed, 55: 97–109.

Styles JA (1978) Mammalian cell transformation in vitro. Br J Cancer, 37: 931–936.

Suzuki J, Koyama T, & Suzuki S (1983) Mutagenicities of mono-nitrobenzene derivatives in the presence of norharman. Mutat Res, 120: 105–110.

Suzuki J, Meguro S, Morita O, Hirayama S, & Suzuki S (1989) Comparison of in vivo binding of aromatic nitro and amino compounds to rat hemoglobin. Biochem Pharmacol, 38: 3511–3519.

Sziza M & Magos L (1959) Toxicologische Untersuchung einiger in der ungarischen Industrie zur Anwendung gelangenden aromatischen Nitroverbindungen. Arch Gewerbepathol Gewerbehyg, 17: 217–226.

Tabak HH, Quave SA, Mashni CI, & Barth EF (1981) Biodegradability studies with organic priority pollutant compounds. J Water Pollut Control Fed, 53: 1503–1518.

Thienes CH & Haley TJ (1955) Clinical toxicology, 3rd ed. London, Henry Kimpton, p 238.

Tonogai Y, Ogawa S, Ito Y, & Iwaida M (1982) Actual survey on TLm (median tolerance limit) values of environmental pollutants, especially on amines, nitriles, aromatic nitrogen compounds and artificial dyes. J Toxicol Sci, **7**: 193–203.

Trabalka JR & Garten CT Jr (1982) Development of predictive models for xenobiotic bio-accumulation in terrestrial systems. Oak Ridge, Tennessee, Oak Ridge National Laboratory, Environmental Sciences Division, 249 pp (ORNL/TM-5869).

Trush MA & Kensler TW (1991) An overview of the relationship between oxidative stress and chemical carcinogenesis. Free Radical Biol Med, **10**: 201–209.

Turney GL & Goerlitz DF (1990) Organic contamination of groundwater at Gas Works Park, Seattle, Washington. Ground Water Monit Rev, **10**: 187–198.

Tyl RW, France KA, Fisher LC, Dodd DE, Pritts IM, Lyon JP, O'Neal FO, & Kimmerle G (1987) Developmental toxicity evaluation of inhaled nitrobenzene in CD rats. Fundam Appl Toxicol, **8**: 482–492.

Urano K & Kato Z (1986) Evaluation of biodegradation ranks of priority organic compounds. J Hazard Mater, **13**: 147–159.

US EPA (1980) Ambient water quality criteria for nitrobenzene. Washington, DC, US Environmental Protection Agency (Report No. EPA 440/5-80-061; NTIS No. PB81-117723).

US EPA (1982a) Test method: Nitroaromatics and isophorone — method 609. In: Longbottom JE & Lichtenberg JJ ed. Test methods: Methods for organic chemical analysis of municipal and industrial wastewater. Washington, DC, US Environmental Protection Agency, Environmental Monitoring and Support Laboratory (Report No. EPA 600/4-82-057).

US EPA (1982b) Test method: Base/neutrals and acids — method 625. In: Longbottom JE & Lichtenberg JJ ed. Test methods: Methods for organic chemical analysis of municipal and industrial wastewater. Washington, DC, US Environmental Protection Agency, Environmental Monitoring and Support Laboratory (Report No. EPA 600/4-82-057).

US EPA (1983) Nitrobenzene. In: Treatability manual. Vol I. Treatability data. Washington, DC, US Environmental Protection Agency, Office of Research and Development (Report No. EPA-600/2-82-001a).

US EPA (1984) US Environmental Protection Agency. Fed Regist, **49**: 25013–25017.

US EPA (1985) Health and environmental effects profile for nitrobenzene. Cincinnati, Ohio, US Environmental Protection Agency, Office of Research and Development (Report No. EPA 600/X-85/365; NTIS No. PB88-180500).

US EPA (1986a) Method 8410: Capillary column analysis of semivolatile organic compounds by gas chromatography/Fourier transform infrared (GC/FT-IR) spectrometry. In: Test methods for evaluating solid waste, 3rd ed. Washington, DC, US Environmental Protection Agency, Office of Solid Waste and Emergency Response (Report SW-846).

US EPA (1986b) Method 8090: Nitroaromatics and cyclic ketones. In: Test methods for evaluating solid waste, 3rd ed. Washington, DC, US Environmental Protection Agency, Office of Solid Waste and Emergency Response (Report SW-846).

US EPA (1986c) Method 8250: Gas chromatographic/mass spectrometry for semivolatile organics: Packed column technique. In: Test methods for evaluating solid waste, 3rd ed. Washington, DC, US Environmental Protection Agency, Office of Solid Waste and Emergency Response (Report SW-846).

US EPA (1986d) Method 8270: Gas chromatographic/mass spectrometry for semivolatile organics: Capillary column technique. In: Test methods for evaluating solid waste, 3rd ed. Washington, DC, US Environmental Protection Agency, Office of Solid Waste and Emergency Response (Report SW-846).

US EPA (1988) US Environmental Protection Agency: Part II. Fed Regist, **53**: 31138–31222.

US EPA (1989) US Environmental Protection Agency: Part II. Fed Regist, **54**: 1056–1119.

US EPA (1991) Alpha-2u-globulin: Association with chemically induced renal toxicity and neoplasia in the male rat. Washington, DC, US Environmental Protection Agency, Risk Assessment Forum (Report No. EPA/625/3-91/019F).

US EPA (1996) Proposed guidelines for cancer risk assessment. Fed Regist, **61**: 17960–18011.

Vance WA & Levin DE (1984) Structural features of nitroaromatics that determine mutagenic activity in *Salmonella typhimurium*. Environ Mutagen, **6**: 797–811.

Van Zoest R & Van Eck GTM (1991) Occurrence and behaviour of several groups of organic micropollutants in the Scheldt Estuary. Sci Total Environ, **103**: 57–71.

Veith GD, DeFoe DL, & Bergstedt BV (1979) Measuring and estimating the bioconcentration factor of chemicals in fish. J Fish Res Board Can, **36**: 1040–1048.

Verna L, Whysner J, & Williams GM (1996) 2-Acetylaminofluorene mechanistic data and risk assessment: DNA reactivity, enhanced cell proliferation and tumor initiation. Pharmacol Ther, **71**: 83–105.

Verschueren K (1983) Handbook of environmental data on organic chemicals. New York, Van Nostrand Reinhold Company.

Volskay VT Jr & Grady CPL Jr (1988) Toxicity of selected RPA compounds to activated sludge microorganisms. J Water Pollut Control Fed, **60**: 1850–1856.

Von Oettingen WF (1941) The aromatic amino and nitro compounds, their toxicity and potential dangers: A review of the literature. Washington, DC, US Public Health Service (Public Health Bulletin No. 271).

Waggot A & Wheatland AB (1978) Contribution of different sources to contamination of surface waters with specific persistent organic pollutants. In: Hutzinger O, Van Leylyveld IH, & Zoeteman BCJ ed. Aquatic pollutants: Transformation and biological effects. Proceedings of the 2nd International Symposium on Aquatic Pollutants, Noordwijkerhout. Amsterdam, Pergamon Press.

Walterskirchen L (1939) Ein Fall von Mirbanölvergiftung.Wein Klin Wochenschr, **52**: 317–318.

Walton BT, Hendricks MS, Anderson TA, & Talmage SS (1989) Treatability of hazardous chemicals in soils: Volatile and semivolatile organics. Oak Ridge, Tennessee, Oak Ridge National Laboratory (US Department of Energy Contract AC05-840R21400; NTIS Order No. DE89016892/XAD).

Walton BT, Hendricks MS, Anderson TA, Griest WH, Merriweather R, Beauchamp JJ, & Francis CW (1992) Soil sorption of volatile and semivolatile organic compounds in a mixture. J Environ Qual, **21**: 552–558.

Wang Y, Wang Z, Liu J, Ma M, & Belzile N (1999) Monitoring priority pollutants in the Yanghe River by dichloromethane extraction and semipermeable membrane device (SPMD). Chemosphere, **39**: 113–131.

Wangenheim J & Bolcsfoldi G (1988) Mouse lymphoma L5178Y thymidine kinase locus assay of 50 compounds. Mutagenesis, 3: 193–205.

Warne MS (1998) Critical review of methods to derive water quality guidelines for toxicants and a proposal for a new framework. Canberra, Environment Australia (Supervising Scientist Report 135).

Warner HP, Cohen JM, & Ireland JC (1987) Determination of Henry's law constants of selected priority pollutants. Washington, DC, US Environmental Protection Agency, pp 1–14 (Report No. EPA/600/D-87/229) [cited in BUA, 1994].

Weant GE & McCormick GS (1984) Nonindustrial sources of potentially toxic substances and their applicability to source apportionment methods. Washington, DC, US Environmental Protection Agency (Report No. EPA-450/4-84-003).

Webber MD & Lesage S (1989) Organic contaminants in Canadian municipal sludges. Waste Manage Res, **7**: 63–82.

Weil C (1972) Statistics vs. safety factors and scientific judgment in the evaluation of safety for man. Toxicol Appl Pharmacol, **21**: 454–463.

Weisberger JH & Weisberger EK (1973) Biochemical formation and pharmacological, toxicological, and pathological properties of hydroxylamines and hydroxamic acids. Pharmacol Rev, **25**: 1–66.

Wellens H (1982) Vergleich der empfindlichkeit von *Brachydario rerio* und *Leuciscus idus* bei der Untersuchung der Fishtoxizitaet von chemischen Verbindungen und Abwassern. Wasser Abwasser Forsch, **15**: 49–52.

Wheeler LA, Soderberg FB, & Goldman P (1975) The relationship between nitro group reduction and the intestinal microflora. J Pharmacol Exp Ther, **194**: 135–144.

White RE (1980) Organic chemical manufacturing 1. Program. Washington, DC, US Environmental Protection Agency (Report No. EPA-450/3-80-023).

WHO (1986) Diseases caused by toxic nitro and amino derivatives of benzene and its homologues. In: Early detection of occupational diseases, Chapter 9. Geneva, World Health Organization.

WHO (1996) Guidelines for drinking-water quality, 2nd ed. Geneva, World Health Organization.

Wild D, Eckhardt K, Gocke E, & King MT (1980) Comparative results of short-term *in vitro* and *in vivo* mutagenicity tests obtained with selected environmental chemicals. In: Norpoth KH ed. Short-term test systems for detecting carcinogens. New York, Springer-Verlag, p 170 [cited in Beauchamp et al., 1982].

Wilmer JL, Kligerman AD, & Erexson GL (1981) Sister chromatid exchange induction and cell cycle inhibition by aniline and its metabolites in human fibroblasts. Environ Mutagen, **3**: 627–638.

Wilson JT, Enfield CG, Dunlap WJ, Cosby RL, Foster DA, & Baskin LB (1981) Transport and fate of selected organic pollutants in a sandy soil. J Environ Qual, **10**: 501–506.

Wirth PJ, Dybing E, von Bahr C, & Thorgeirsson SS (1980) Mechanism of *N*-hydroxyacetylaryl-amine mutagenicity in the *Salmonella* test system: Metabolic activation of *N*-hydroxyphenacetin by liver and kidney fractions from rat, mouse, hamster, and man. Mol Pharmacol, **18**: 117–127.

Wirtschafter ZT & Wolpaw R (1944) A case of nitrobenzene poisoning. Ann Intern Med, **21**: 135–140.

Wiśniewska-Knypl JM, Jabłońska JK, & Piotrowski JK (1975) Effect of repeated exposure to aniline, nitrobenzene, and benzene on liver microsomal metabolism in the rat. Br J Ind Med, **32**: 42–48.

Witte F & Zetzsch C (1984) The temperature dependence of the forward-backward reactions of the addition of OH to benzene, aniline and nitrobenzene. In: Physical-chemical behaviour of atmospheric pollutants. Commission of the European Communities, pp 168–176 ((REP) EUR 9436).

Witte F, Urbanik E, & Zetzsch C (1986) Temperature dependence of the rate constants for the addition of OH to benzene and to some monosubstituted aromatics (aniline, bromobenzene, and nitrobenzene) and the unimolecular decay of the adducts. Kinetics into a quasi-equilibrium. 2. J Phys Chem, **90**: 3251–3259.

Wolfe NL (1992) Abiotic transformations of pesticides in natural waters and sediments. In: Schnoor JR ed. Fate of pesticides and chemicals in the environment. Chichester, John Wiley and Sons.

Wright-Smith RJ (1929) Poisoning by nitrobenzene or "essence of mirbane" with recovery. Med J Aust, **1**: 867–868.

Wujcik WJ, Lowe WL, Marks PJ, & Sisk WE (1992) Granular activated carbon pilot treatment studies for explosives removal from contaminated groundwater. Environ Prog, **11**: 178–189.

Yamamoto T, Yoneyama M, Imanishi M, & Takeuchi M (2000) Flow cytometric detection and analysis of tailless sperm caused by sonication or a chemical agent. J Toxicol Sci, **25**: 41–48.

Yoshida K, Shigeoka T, & Yamauchi F (1988) Estimation of environmental fate of industrial chemicals. Toxicol Environ Chem, **17**: 69–85.

Yoshida T (1962) Experimental studies on the histopathological changes of the optic nerve of rabbits administered with aniline or nitrobenzene. Jpn J Ind Health, **4**: 262–280.

Yoshioka Y, Ose Y, & Sato T (1985) Testing for the toxicity of chemicals with *Tetrahymena pyriformis*. Sci Total Environ, **43**: 149–157.

Yoshioka Y, Nagase H, Ose Y, & Sato T (1986) Evaluation of the test method "activated sludge, respiration inhibition test" proposed by the OECD. Ecotoxicol Environ Saf, **12**: 206–212.

Yoshioka T, Suzuki T, & Uematsu T (1989) Biotransformation of *N*-substituted aromatic compounds in mammalian spermatozoa. Nonoxidative formation of *N*-hydroxy-*N*-aryl-acetamides from nitroso aromatic compounds. J Biol Chem, **264**: 12432–12438.

Young DR, Gossett RW, Baird RB, Brown DA, Taylor PA, & Miille MJ (1983) Wastewater inputs and marine bioaccumulation of priority organic pollutants off southern California. In: Jolley RL, Brungs WA, & Cotruvo JA ed. Water chlorination: Environmental impact and health effects. Vol 4, Book 2: Environmental health and risk; Proceedings of the 4th Conference. Ann Arbor, Michigan, Ann Arbor Science Publishers, pp 871–884.

Yu YS & Bailey GW (1992) Soil process and chemical transport. Reduction of nitrobenzene by four sulfide minerals: kinetics, products, and solubility. J Environ Qual, **21**: 86–94.

Zeise L, Wilson R, & Crouch E (1984) Use of acute toxicity to estimate carcinogenic risk. Risk Anal, **4**: 187–199.

Zeise L, Crouch E, & Wilson R (1986) A possible relationship between toxicity and carcinogenicity. J Am Coll Toxicol, **5**: 137–151.

Zeitoun MM (1959) Nitrobenzene poisoning in infants due to inunction with false bitter almond oil. J Trop Pediatr, **5**: 73–75.

Zeligs M (1929) Aniline and nitrobenzene poisoning in infants. Arch Pediatr, **46**: 502–506.

Zepp RG & Schlotzhauer PF (1983) Influence of algae on photolysis rates of chemicals in water. Environ Sci Technol, **17**: 462–468.

Zepp RG, Braun AM, Hoigné J, & Leenheer JA (1987a) Photoproduction of hydrated electrons from natural organic solutes in aquatic environments. Environ Sci Technol, **21**: 485–490.

Zepp RG, Hoigné J, & Bader H (1987b) Nitrate-induced photooxidation of trace organic chemicals in water. Environ Sci Technol, 21: 443–450.

Zoeteman BCJ, Harmsen K, Linders JBHJ, Morra CHF, & Slooff W (1980) Persistent organic pollutants in river water and ground water of the Netherlands. Chemosphere, 9: 231–249.

Zuccola P (1919) Nitrobenzene poisoning. J Am Med Assoc, 72: 231(Abstr).

RESUME

1. Identité, propriétés physiques et chimiques et méthodes d'analyse

Le nitrobenzène se présente sous la forme d'un liquide huileux incolore à jaune pâle, dont l'odeur rappelle celle des amandes amères ou du "cirage." Son point de fusion est de 5,7 °C et son point d'ébullition de 211 °C. A 20 °C, sa tension de vapeur est égale à 20 Pa et sa solubilité dans l'eau à 1900 mg/litre. Avec un point d'éclair (mesuré en coupelle fermée) de 88 °C et une limite inférieure d'explosivité de 1,8 % en volume dans l'air, il y a risque d'incendie en présence de ce produit. Le logarithme de son coefficient de partage entre l'octanol et l'eau est égal à 1,85.

Un certain nombre de méthodes d'analyse sont utilisables pour le dosage du nitrobenzène dans l'air, l'eau et le sol. On dispose également de méthodes pour la surveillance des travailleurs exposés en permanence à ce composé. Il existe aussi une méthode basée sur la chromatographie liquide haute performance sur phases inversées qui convient bien au dosage des métabolites urinaires, et notamment du *p*-nitrophénol (qui est également un métabolite de deux insecticides organophosphorés, le parathion et le parathion-méthyl). D'autres méthodes permettent de doser l'aniline libérée des adduits à l'hémoglobine ou encore la méthémoglobine produite par les métabolites du nitrobenzène.

2. Sources d'exposition humaine et environnementale

Le nitrobenzène n'existe pas à l'état naturel. C'est un produit synthétique utilisé à plus de 95 % pour la préparation de l'aniline - un intermédiaire très important pour la fabrication des polyuréthanes - comme solvant dans le raffinage du pétrole et la fabrication des éthers et des acétates de cellulose, pour la production des dinitrobenzènes et des dichloranilines ou encore pour la synthèse d'autres composés organiques comme l'acétaminophène.

Au début du vingtième siècle, le nitrobenzène était quelque peu utilisé comme additif alimentaire (en tant que substitut de l'essence d'amande amère) et très utilisé comme solvant dans toutes sortes de produits : cirages, encres (utilisées notamment dans les hôpitaux pour tamponner les couches sortant de lessive) et divers désinfectants, de sorte que le risque d'exposition de la population était important à cette époque.

D'après les statistiques dont on dispose, il apparaît que la production annuelle de nitrobenzène a sensiblement augmenté au cours des 30 ou 40 dernières années. La majorité de cette production reste en circuit fermé en vue d'autres synthèses et notamment pour la préparation de l'aniline, mais aussi pour celle des dérivés substitués des nitrobenzènes et de l'aniline. Les pertes de nitrobenzène au cours de la production sont vraisemblablement faibles; toutefois, lorsque ce composé est utilisé comme solvant, les émissions peuvent sans doute être plus importantes. On a montré qu'il pouvait y avoir dégagement de nitrobenzène lors de l'incinération des boues d'égout et sa présence a été décelée en quantités mesurables dans l'air surmontant les décharges de produits dangereux.

Du nitrobenzène peut se former dans l'atmosphère par suite de l'attaque du benzène par les oxydes d'azote, mais cet apport n'a pas été chiffré. Il semblerait également qu'au contact de l'ozone, l'aniline s'oxyde lentement en nitrobenzène.

3. Transport, distribution et transformation dans l'environnement

Le nitrobenzène peut être décomposé par photolyse ou par biodégradation microbienne.

Au vu des propriétés physiques du nitrobenzène, on peut penser qu'il passe en quantité notable de l'eau dans l'atmosphère, sans toutefois qu'il s'agisse d'un processus rapide. Dans l'air et dans l'eau, il subit une lente photodécomposition. L'étude expérimentale directe de la photolyse du nitrobenzène dans l'air a permis de déterminer que sa durée de séjour y était inférieure à 1 jour; en revanche, le calcul de la demi-vie du composé sur la base de sa réaction avec les radicaux hydroxyle donne des valeurs comprises entre 19 et 223 jours. Avec l'ozone, la réaction est encore plus lente. Une étude expérimentale en chambre à brouillard en présence d'un mélange de butane, de dioxyde d'azote et de propylène a permis d'estimer la demi-vie du nitrobenzène dans ces conditions à une valeur comprise entre 4 et 5 jours. Dans les étendues d'eau, la photolyse directe se révèle être le mode de décomposition le plus rapide (demi-vie comprise entre 2,5 et 6 jours), la photolyse indirecte (photo-oxydation par les radicaux hydroxyle, par des électrons hydratés ou en présence d'atomes d'hydrogène ou encore sensibilisation par les acides humiques) ne jouant qu'un rôle minime (la demi-vie calculée se situe dans ce cas entre 125 jours et 13 ans pour la réaction avec les radicaux hydroxyle, en fonction de la concentration de sensibilisateur).

En raison de sa solubilité modérée dans l'eau et de la valeur relativement faible de sa tension de vapeur, on pourrait penser qu'une certaine proportion du nitrobenzène présent dans l'atmosphère est susceptible d'en être éliminée par les précipitations; en fait, l'expérimentation sur le terrain montre que l'élimination du nitrobenzène par les précipitations (soit par dissolution dans les gouttes de pluie, soit par élimination du composé adsorbé sur des particules) ou par dépôt à sec, est négligeable. Il est probable qu'en raison de sa forte densité de sa vapeur (4,1-4,25 fois celle de l'air), une partie du nitrobenzène s'élimine par dépôt de la phase vapeur.

Les données dont on dispose au sujet de l'évaporation du benzène présent dans les étendues d'eau se révèlent quelque peu contradictoires, puisque selon un modèle informatique, la demi-vie de volatilisation irait de 12 jours (cours d'eau) à 68 jours (lac eutrophe) alors que l'estimation la plus courte citée dans la littérature est de 1 jour pour de l'eau de rivière. Une autre étude sur microcosmes expérimentaux dans laquelle on a simulé la décharge au sol d'eaux usées, a montré que le nitrobenzène ne se volatilisait pas, mais qu'il était totalement décomposé.

L'étude de la décomposition du nitrobenzène donne à penser que le composé subit une décomposition aérobie dans les usines de traitement des eaux d'égout, la dégradation étant plus lente en anaérobiose. S'il est présent à forte concentration dans les eaux d'égout, le nitrobenzène ne sera pas forcément totalement décomposé. D'ailleurs, une forte teneur en nitrobenzène est également susceptible d'inhiber la décomposition des autres déchets. La biodégradation du nitrobenzène dépend principalement de l'acclimatation de la population microbienne. La biodégradation par des inoculums non acclimatés est généralement très lente, voire négligeable, et elle ne s'amorce qu'après une longue période d'acclimatation. Les microorganismes acclimatés, notamment ceux qui proviennent des usines de traitement des eaux usées industrielles, se révèlent en revanche capables d'éliminer le composé en l'espace de quelques jours. On constate en général, que cette biodégradation est accrue par la présence d'autres substrats facilement biodégradables. En ce qui concerne la décomposition dans le sol, il semble aussi que les facteurs limitants soient l'adaptation de la microflore et la présence d'autres substrats. On a montré que la décomposition anaérobie du nitrobenzène est très lente, même après une longue période d'acclimatation.

La mesure des facteurs de bioconcentration du nitrobenzène chez un certain nombre d'organismes montre que ce composé n'a qu'un potentiel de bioaccumulation minime et ne subit pas de bioamplification le long de la chaîne alimentaire. Il est possible que les plantes fixent le nitrobenzène; toutefois, selon les résultats dont on dispose, s'il est présent au niveau de l'appareil radiculaire, on ne le retrouve qu'en très faible quantité dans les autres parties de la plante. Dans une expérience de simulation de l'écosystème constitué par une mare de ferme, on a constaté que le nitrobenzène restait en majeure partie dans l'eau et n'était ni accumulé, ni écoamplifié par les microcrustacés, les larves de moustiques, les gastéropodes, les algues ou diverses espèces de plancton et de poissons.

4.   Concentrations dans l'environnement et exposition humaine

La concentration du nitrobenzène dans les prélèvements effectués dans l'environnement - eaux de surface ou souterraines, air, etc. - est généralement faible.

Au début des années 1980 on a relevé dans l'air de villes des Etats-Unis des concentrations allant de < 0,05 à 2,1 $\mu g/m^3$ (< 0,01 et 0,41 partie par milliard) (en moyenne arithmétique). Selon les données publiées en 1985 par l'Environmental Protection Agency des Etats-Unis, moins de 25 % des échantillons d'air prélevés contenaient du nitrobenzène, la concentration médiane étant d'environ 0,05 $\mu g/m^3$ (0,01 partie par milliard); dans les zones urbaines, la concentration était généralement inférieure à 1 $\mu g/m^3$ (0,2 partie par milliard), avec des valeurs légèrement plus élevées dans les zones industrielles (moyenne 2,0 $\mu g/m^3$, soit 0,40 partie par milliard). Sur 49 échantillons d'air analysés au Japon en 1991, 42 se sont révélés contenir du nitrobenzène en quantité mesurable (0,0022-0,16 µg/m^3). Dans l'air au-dessus des zones urbaines et des sites de décharge, la concentration était sensiblement plus faible (ou même non décelable) en hiver qu'en été.

On possède davantage de données sur la teneur des eaux de surface en nitrobenzène que sur les concentrations atmosphériques. Les valeurs varient selon le lieu et la saison, mais elles sont généralement faibles (environ 0,1 à 1 µg/litre). L'une des valeurs les plus fortes (67 µg/litre) a été relevée en 1990 en Yougoslavie, dans les eaux du Danube. Aux Etats-Unis en revanche, on n'a décelé de nitrobenzène dans aucun des échantillons d'eaux de surface prélevés à proximité de nombreuses décharges de produits dangereux (statistiques de 1988). En s'appuyant sur des données en nombre limité, on peut dire que le risque de contamination est plus élevé pour les eaux souterraines que pour les eaux de surface; des prélèvements effectués en plusieurs points du territoire des Etats-Unis vers la fin des années 1980 ont révélé des concentrations 210 à 250, voire 1400 µg/litre (avec des valeurs encore beaucoup plus élevées pour un site de gazéification de la houille). Des analyses effectuées sur de l'eau de boisson aux Etats-Unis et au Royaume-Uni au cours des années 1970 et 1980, ont révélé la présence de nitrobenzène, mais seulement dans une faible proportion des échantillons. Au Canada, par contre, sur 30 échantillons analysés, aucun ne contenait du nitrobenzène (statistiques de 1982).

On n'a pas trouvé de données relatives à la présence de nitrobenzène dans les denrées alimentaires, encore que des études effectuées au Japon en 1991 aient permis d'en découvrir dans une petite fraction (4 sur 147) des échantillons de poisson analysés. Une étude menée en 1985 aux Etats-Unis sur des échantillons de biotes très divers, n'a pas permis d'en déceler la présence.

La population dans son ensemble est susceptible d'être exposée au nitrobenzène présent dans l'air ou éventuellement dans l'eau de boisson. Il y a également un risque d'exposition dû à certains produits de consommation, mais on manque de données précises à ce sujet. Selon des études effectuées dans l'Etat du New Jersey, sur la côte est des Etats-Unis (qui connaît des étés chauds à torrides et des hivers froids à glacials), les concentrations sont plus élevées en été qu'en hiver du fait de la formation de nitrobenzène par nitration du benzène provenant de l'essence et de la plus grande volatilité du nitrobenzène pendant les grandes chaleurs; en hiver, l'exposition par l'intermédiaire de l'air ambiant est sans doute négligeable. Si l'on se base sur les analyses de l'air et l'estimation des émissions au cours de la production, seules les populations qui résident à proximité d'installations industrielles (c'est-à-dire d'unités de production ou d'unités utilisant du nitrobenzène comme intermédiaire de synthèse) ou de raffineries de pétrole, sont susceptibles d'être sensiblement exposées au nitrobenzène. Cependant, les personnes qui vivent sur d'anciennes décharges de déchets dangereux ou aux alentours de tels sites risquent également une exposition plus importante, soit du fait de la pollution des eaux souterraines, soit en raison de la contamination du sol et de la fixation du composé par les plantes.

Les niveaux d'exposition professionnelle devraient être inférieurs à la limite généralement adoptée pour l'exposition par voie aérienne (5 mg/m^3, soit 1 ppm). Les données disponibles montrent que le nitrobenzène est bien absorbé par la voie transcutanée, sous la forme de vapeur ou de liquide; l'exposition cutanée pourrait donc être importante, mais on manque de données sur ce point.

5. Cinétique et métabolisme

Le nitrobenzène est un liquide volatil qui peut facilement pénétrer dans l'organisme par inhalation, par voie percutanée sous forme de vapeur, ou encore par ingestion ou pénétration du liquide dans le derme. Chez le rat, l'activation du nitrobenzène en métabolites conduisant à la formation de méthémoglobine se révèle dépendre en grande partie de la flore intestinale. L'expérimentation animale montre que la majeure partie du nitrobenzène (environ 80 % de la dose) est métabolisée et éliminée en l'espace de 3 jours. Le reste n'est que lentement éliminé. Cette lenteur est probablement due au recyclage érythrocytaire des formes rédox du composé et des conjugués avec le glutathion. On a mis en évidence la formation de liaisons covalentes, vraisemblablement avec les groupements sulfhydryle de l'hémoglobine.

Chez les rongeurs et les lapins, les principaux métabolites urinaires sont constitués par le *p*-nitrophénol et le *p*-aminophénol. Chez l'être humain, une partie de la dose absorbée est excrétée par la voie urinaire; 10 à 20 % de cette dose sont excrétés sous forme de *p*-nitrophénol (qui peut donc être utilisé pour la surveillance biologique).

La demi-vie d'élimination est estimée à environ 5 h pour le *p*-nitrophénol pendant la phase initiale et à plus de 20 h pendant la phase finale. L'autre métabolite urinaire (le *p*-aminophénol) n'est présent en quantité importante qu'en cas d'exposition à une dose élevée.

6. Effets sur les mammifères de laboratoire et les systèmes d'épreuve *in vitro*

Le nitrobenzène exerce ses effets toxiques sur de multiples organes et par toutes les voies d'exposition. Chez le rat et la souris, l'exposition par voie orale, cutanée, sous-cutanée ou respiratoire provoque une méthémoglobinémie qui a pour conséquence une anémie hémolytique, une congestion de la rate et du foie ainsi qu'une hémopoïèse médullaire et splénique.

Des lésions capsulaires de la rate ont été observées chez des rats ayant reçu du nitrobenzène par gavage (à des doses quotidiennes ne dépassant pas 18,75 mg/kg de poids corporel) ou par application cutanée (à raison de 100 mg/kg p.c. par jour et au-delà). Des lésions spléniques analogues ont déjà été observées après exposition à des colorants à base d'aniline, qui pour certains d'entre eux, ont provoqué des sarcomes de la rate lors d'études de cancérogénicité chronique effectués sur des rats. Après administration par gavage ou application cutanée, on a observé chez des rats et des souris la présence de lésions hépatiques consistant en une nécrose des hépatocytes centrolobulaires, la présence de nucléoles proéminents dans les hépatocytes, une dégénérescence hydropique sévère et une accumulation de pigments dans les cellules de Kupffer. Une augmentation de la vacuolisation de la zone X de la surrénale a été observée chez des souris femelles après administration par voie orale et application cutanée.

Lors d'études subchroniques au cours desquelles le composé a été administré à des souris et à des rats soit par voie orale, soit en applications cutanées, on a observé, au niveau du cervelet et du tronc cérébral, des lésions du système nerveux central engageant le pronostic vital. Ces lésions, qui consistaient également en pétéchies hémorragiques, pourraient être soit des effets toxiques directs, soit des effets indirects dus à l'hypoxie ou à l'hépatotoxicité du nitrobenzène. Selon la dose administrée, ces effets neurotoxiques se sont manifestés cliniquement par une ataxie, une tête penchée et arquée, la perte du réflexe de redressement, des tremblements, un coma et des convulsions.

Parmi les autres organes cibles figure le rein (poids augmenté, gonflement de l'épithélium glomérulaire et tubulaire, pigmentation des cellules de l'épithélium tubulaire), l'épithélium nasal (glandularisation de l'épithélium respiratoire, dépôt de pigments dans l'épithélium olfactif, avec dégénérescence épithéliale), la thyroïde (hyperplasie folliculaire), le thymus (involution) et le pancréas (infiltration par des mononucléaires). Chez le lapin, on observe des lésions anatomopathologiques au niveau pulmonaire (emphysème, atelectasie et bronchiolisation de la paroi alvéolaire).

L'exposition prolongée (505 jours) de souris B6C3F$_1$ mâles et femelles, ainsi que de rats et de rattes Fischer-344 et Sprague-Dawley, a permis d'évaluer la toxicité et la cancérogénicité potentielles du nitrobenzène administré par la voie respiratoire. Il n'y a pas eu d'effet négatif sur la survie aux concentrations utilisées (jusqu'à 260 mg/m^3, soit 50 ppm pour la souris; jusqu'à 130 mg/m^3, soit 25 ppm pour les rats), mais l'inhalation de nitrobenzène s'est révélée toxique et cancérogène chez les deux espèces animales et les deux souches de rats, provoquant un éventail de tumeurs bénignes ou malignes du poumon, de la thyroïde, de la glande mammaire, du foie ou du rein.

Le nitrobenzène s'est révélé dépourvu de génotoxicité *in vitro* pour les cellules bactériennes et mammaliennes et *in vivo* pour les cellules mammaliennes. Les études publiées ont notamment porté sur l'endommagement et la réparation de l'ADN, les mutations géniques, les effets chromosomiques et la transformation cellulaire.

De nombreux travaux confirment la toxicité testiculaire du nitrobenzène, les points d'aboutissement les plus sensibles de cette action toxique étant, dans l'ordre, la mobilité et le nombre de spermatozoïdes, puis leur mobilité progressive et leur viabilité ainsi que la présence de spermatozoïdes anormaux et enfin, la diminution de l'indice de fécondité.

Lors d'une étude sur la toxicité génésique (toxicité pour la fonction de reproduction) du nitrobenzène au cours de laquelle on a fait inhaler le produit à des rats Sprague-Dawley, on a constaté qu'à la concentration de 200 mg/m^3 (40 ppm) mais à l'exclusion de celles de 5 et de 51 mg/m^3 (1 et 10 ppm), l'indice de fécondité des générations

F_0 et F_1 était fortement réduit, en relation avec des effets toxiques sur l'appareil reproducteur mâle; cette hypofertilité s'est révélée partiellement réversible, lorsque les animaux de la génération F_1 soumis à la concentration de 200 mg/m³ ont été accouplés avec des femelles vierges non traitées, après une période de récupération de 9 semaines. Toutefois, une étude effectuée sur la même souche de rats, mais en utilisant la voie orale cette fois (20-100 mg/kg de poids corporel à partir du 14ème jour précédant l'accouplement jusqu'au 4ème jour de lactation), a montré qu'à part une réduction du poids des ratons et une augmentation de la mortalité postnatale, le nitrobenzène n'avait eu aucun effet sur les paramètres génésiques. L'absence d'effet sur la fécondité dans cette étude s'explique par le court intervalle de temps entre l'administration et l'accouplement et également par le fait que les rats produisent des spermatozoïdes en très grand excès. Une réduction de la fertilité des mâles associée à une importante atrophie testiculaire a été observée chez des rats et des souris; chez la souris, les effets se sont manifestés à la dose quotidienne de 300 mg/kg p.c. administrée par gavage et à la dose de 800 mg/kg p.c. administrée quotidiennement par application cutanée. Chez le rat, ces effets ont fait leur apparition à la dose quotidienne de 75 mg/kg p.c. administrée par gavage et de 400 mg/kg p.c. administrée quotidiennement par application cutanée. Les manifestations toxiques au niveau testiculaire consistaient en une desquamation de l'épithélium séminifère, l'apparition de cellules géantes polynucléées, une atrophie macroscopique et une aspermie prolongée. Les études *in vivo* et *in vitro* montrent que le nitrobenzène a des effets directs sur le testicule. Ces effets affectent la spermatogénèse, avec une exfoliation qui touche des cellules germinales pour la plupart viables et une dégénérescence des cellules de Sertoli. Le principal effet histopathologique consiste dans la dégénérescence des spermatocytes.

En règle générale, on n'a pas constaté d'atteinte au niveau des organes reproducteurs femelles, sauf dans le cas d'une étude qui a mis en évidence une atrophie de l'utérus chez la souris après administration quotidienne d'une dose de 800 mg/kg p.c. par application cutanée.

En ce qui concerne les effets toxiques sur le développement, les études effectuées sur des rats et des lapins indiquent que l'exposition au nitrobenzène par la voie respiratoire ne produit pas d'effets foetotoxiques, embryotoxiques ou tératogènes aux concentrations qui sont toxiques pour la mère. A la concentration la plus forte étudiée

(530 mg/m³, soit 104 ppm, chez le lapin), le nombre moyen de sites de résorption et la rapport résorptions/nidations étaient plus nombreux chez les animaux traités que chez les témoins concomitants mais ils se situaient dans la limite des valeurs relevées chez les témoins historiques. Des effets (augmentation du taux de méthémoglobine et du poids du foie) ont été observés sur les mères à la concentration de 210 mg/m³ (41 ppm).

Une étude sur l'immunotoxicité du nitrobenzène effectuée avec des souris B6C3F$_1$ a mis en évidence une augmentation de la cellularité de la rate, une certaine immunodépression se traduisant par une réduction de la réponse en IgM aux érythrocytes de mouton et une stimulation de la moelle osseuse. La résistance de l'hôte aux infections microbiennes ou virales n'a pas sensiblement souffert de l'administration de nitrobenzène, si ce n'est une tendance à l'augmentation de la sensibilité chez les cas pour lesquelles les cellules T intervenaient dans les mécanismes de défense.

7. Effets sur l'Homme

Le nitrobenzène est toxique pour l'être humain en cas d'exposition par voie cutanée, respiratoire ou orale. Chez l'Homme, le principal effet général consécutif à une exposition au nitrobenzène est une méthémoglobinémie.

On connaît de nombreux cas d'intoxication et de décès dus à l'ingestion de nitrobenzène. En cas d'ingestion ou lorsque le malade était semble-t-il sur le point de mourir par suite d'une méthémoglobinémie sévère, la cessation de l'exposition et une prompte intervention médicale lui ont permis de se rétablir progressivement jusqu'à récupération complète. Une dose de nitrobenzène suffisamment forte peut être mortelle quelle que soit la voie d'absorption, mais on estime peu probable la survenue d'une exposition assez intense pour avoir une issue fatale, sauf en cas d'accident industriel ou de suicide.

En cas d'exposition humaine au nitrobenzène, c'est probablement la rate qui constitue un des organes cibles. Ainsi, chez une femme exposée de par sa profession à du nitrobenzène présent dans de la peinture, on a constaté une splénomégalie accompagnée d'une sensibilité de la rate à la palpation.

Chez une autre femme exposée à du nitrobenzène par la voie respiratoire, on a constaté des effets hépatiques consistant notamment en hépatomégalie avec sensibilité à la palpation ainsi que des anomalies dans les paramètres biochimiques sanguins.

Les symptômes de neurotoxicité observés chez l'Homme après inhalation de nitrobenzène, consistent en céphalées, confusion, vertiges et nausées. Ces symptômes s'observent également après ingestion de nitrobenzène, avec en outre apnée et coma.

8.   Effets sur les êtres vivants dans leur milieu naturel

Le nitrobenzène se révèle toxique pour les bactéries et il peut gêner le fonctionnement des installations de traitement des eaux d'égout lorsqu'il y est présent à forte concentration. C'est pour la bactérie *Nitrosomonas* que l'on a trouvé la concentration toxique la plus faible pour un microorganisme, avec une CE_{50} (concentration qui provoque 50 % d'inhibition d'une activité ou d'un processus - dans le cas présent l'inhibition de la consommation d'ammoniac) égale à 0,92 mg/litre. En ce qui concerne les autres valeurs, on a 1,9 mg/litre pour la concentration sans effet observable (NOEC) à 72 h sur le protozoaire *Entosiphon sulcatum*, et également 1,9 mg/litre pour la concentration la plus faible produisant un effet à 8 jours (LOEC) sur l'algue bleue *Microcystis aeruginosa*.

En ce qui concerne les invertébrés d'eau douce, les valeurs de la toxicité aiguë (CL_{50} à 24-48 h) vont de 24 mg/litre pour la daphnie (*Daphnia magna*) à 140 mg/litre pour le gastéropode *Lymnaea stagnalis*. S'agissant des invertébrés marins, on donne comme valeur la plus faible le chiffre de 6,7 mg/litre pour la CL_{50} à 96 h dans le cas d'une crevette de la famille des mysidés, *Mysidopsis bahia*. La valeur la plus faible obtenue dans un test de toxicité chronique est de 1,9 mg/litre pour la NOEC à 20 jours et elle concerne *Daphnia magna*. Pour ce même invertébré, la CE_{50} relative à la reproduction est de 10 mg/litre.

Les poissons d'eau douce sont également peu sensibles au nitrobenzène. On a obtenu, pour la CL_{50} à 96 h, des valeurs qui vont de 24 mg/litre dans le cas du medaka (*Oryzias latipes*) à 142 mg/litre dans le cas du guppy (*Poecilia reticulata*). Aucun effet n'a été constaté sur la mortalité ou le comportement du medaka lors d'une exposition de 18 jours à la concentration de 7,6 mg/litre.

9. Evaluation du danger et du risque

Dans les cas d'exposition humaine, on observe une méthémoglobinémie suivie d'anomalies hématologiques et spléniques, mais les données disponibles ne permettent pas de déterminer une relation exposition-réponse quantitative. Chez les rongeurs, on constate aussi une méthémoglobinémie ainsi que des effets hématologiques et des effets sur le testicule auxquels s'ajoutent également des effets sur l'appareil respiratoire révélés par les tests d'inhalation. Une méthémoglobinémie, une hypospermie épididymaire bilatérale et une atrophie testiculaire également bilatérale ont été observées chez le rat à la concentration la plus faible étudiée (5 mg/m^3, soit 1 ppm). Chez la souris, on a constaté une bronchiolisation plus fréquente des parois alvéolaires ainsi qu'une hyperplasie broncho-alvéolaire à la concentration la plus faible étudiée (26 mg/m^3, soit 5 ppm). Une réaction cancérogène a été observée chez le rat et la souris après exposition au nitrobenzène : il s'agissait de carcinomes mammaires chez des souris femelles B6C3F$_1$, de carcinomes hépatiques chez des rats mâles Fischer-344 ou d'adénocarcinomes folliculaires de la thyroïde chez des rats mâles de cette même souche. Des tumeurs bénignes ont été observées dans cinq organes. Les études de génotoxicité ont généralement donné des résultats négatifs.

Bien que plusieurs métabolites du nitrobenzène puissent être considérés comme étant à l'origine des propriétés cancérogènes de ce composé, on ignore quel en est le mécanisme. Les mécanismes rédox étant sans doute communs à l'être humain et aux animaux de laboratoire, on suppose que le nitrobenzène est susceptible de provoquer des cancers chez l'Homme quelle que soit la voie d'exposition.

Il est probable que la population générale est peu exposée au nitrobenzène par l'air ou l'eau. Il n'a été possible de tirer une valeur de la dose sans effet nocif observable (NOAEL) d'aucune des études toxicologiques effectuées, mais il semble que le risque d'effets non néoplasiques soit faible. Si l'exposition est suffisamment faible pour ne pas provoquer de tels effets, il est également probable qu'il n'y aura pas non plus d'effets cancérogènes.

Des intoxications par le nitrobenzène présent dans divers produits de consommation se sont fréquemment produites par le passé. Comme la tension de vapeur du composé est modérément forte et qu'il est largement résorbé par la voie percutanée, une exposition humaine importante est possible par cette voie. En outre, le produit dégage une odeur d'amande amère assez agréable, de sorte qu'on n'est pas forcément dissuadé de consommer des aliments ou de l'eau contaminés. Les nourrissons sont particulièrement sensibles aux effets du nitrobenzène.

Les données relatives à l'exposition sur le lieu de travail sont limitées. Une étude effectuée dans ces conditions a montré que les travailleurs étaient exposés à des concentrations du même ordre que la LOAEL (concentration la plus faible provoquant un effet nocif observable) déterminée à l'occasion d'une étude avec exposition prolongée par la voie respiratoire. Il y a donc là un sérieux motif d'inquiétude pour la santé des travailleurs exposés au nitrobenzène.

Le nitrobenzène n'a guère tendance à la bioaccumulation et il peut subir une biotransformation par voie aérobie ou anaérobie. En ce qui concerne les organismes terrestres, il y a peu de chances qu'on retrouve dans l'environnement le même degré de gravité que dans les tests de laboratoire, sauf peut-être dans des zones proches d'entreprises où l'on produit ou utilise du nitrobenzène ou qui ont pu être contaminées par un déversement accidentel.

En utilisant d'une part, les données de toxicité aiguë disponibles et une distribution statistique, et, d'autre part, le rapport de la toxicité aiguë à la toxicité chronique tiré des résultats obtenus sur des crustacés, on peut estimer à 200 µg/litre la limite de concentration de nitrobenzène en-deçà de laquelle 95 % des espèces dulçaquicoles sont protégées avec un degré de confiance de 50 %. Il est donc peu probable que le nitrobenzène présent dans l'environnement représente un danger pour les espèces aquatiques, compte tenu des concentrations habituellement mesurées dans les eaux de surface, à savoir entre 0,1 et 1 µg/litre. Même à la concentration la plus élevée qui ait été signalée (67 µg/litre), le nitrobenzène ne devrait pas constituer un sujet de préoccupation pour les organismes d'eau douce.

Les informations disponibles sont insuffisantes pour que l'on puisse établir une valeur-guide applicable aux organismes marins.

RESUMEN

1. Identidad, propiedades físicas y químicas y métodos analíticos

El nitrobenceno es un líquido oleoso de incoloro a amarillo pálido con un olor que recuerda al de las almendras amargas o el "betún para el calzado." Tiene un punto de fusión de 5,7°C y un punto de ebullición de 211°C. Su presión de vapor es de 20 Pa a 20°C y su solubilidad en agua de 1900 mg/l a 20°C. Representa un peligro de incendio, con un punto de inflamación (método del vaso cerrado) de 88°C y un límite explosivo (inferior) de 1,8% en volumen en el aire. Su log del coeficiente de reparto octanol/agua es de 1,85.

Hay una serie de métodos analíticos para la cuantificación del nitrobenceno en muestras de aire, agua y suelo. También existen métodos para la vigilancia de los trabajadores expuestos sistemáticamente a esta sustancia. Para la determinación de metabolitos urinarios, incluido el p-nitrofenol (también metabolito urinario de los insecticidas órganofosforados paratión y metilparatión), parece adecuado el método de la cromatografía líquida de alto rendimiento en fase invertida. También hay métodos disponibles para la determinación de la anilina liberada a partir de los aductos de hemoglobina y para determinar la metahemoglobina que producen los metabolitos del nitrobenceno.

2. Fuentes de exposición humana y ambiental

El nitrobenceno no se encuentra de forma natural en el medio ambiente. Es un producto sintético, más del 95% del cual se utiliza en la producción de anilina, intermediario químico importante que se usa en la fabricación de poliuretanos; el nitrobenceno se usa también como disolvente en el refinado del petróleo, como disolvente en la fabricación de éteres y acetatos de celulosa, en la fabricación de dinitrobencenos y dicloroanilinas y en la síntesis de otros compuestos orgánicos, incluido el acetaminofén.

A comienzos del siglo XX, el nitrobenceno se utilizaba algo como aditivo de los alimentos (sustitutivo de la esencia de almendras) y también de manera generalizada como disolvente en diversos productos patentados, en particular betún para calzado, tintas (incluidas las tintas utilizadas para estampar los pañales de bebé recién lavados de los hospitales) y en varios desinfectantes, de manera que entonces se había un potencial significativo de exposición del público.

Los registros disponibles ponen de manifiesto que ha habido un aumento significativo de la producción anual de nitrobenceno durante los 30-40 últimos años. La mayor parte se mantiene en sistemas cerrados para utilizarlo en otras síntesis, particularmente de anilina, pero también de nitrobencenos y anilinas sustituidos. Las pérdidas durante la producción de nitrobenceno probablemente son bajas; sin embargo, cuando el nitrobenceno se utiliza como disolvente las emisiones pueden ser más altas. Se ha demostrado que las instalaciones de incineración de fangos cloacales emiten nitrobenceno y se ha medido en el aire en los vertederos de residuos peligrosos.

Se puede formar nitrobenceno en la reacción del benceno en la atmósfera en presencia de óxidos de nitrógeno, aunque no se ha cuantificado esta fuente. Se ha notificado que la anilina se oxida lentamente a nitrobenceno mediante el ozono.

3. Transporte, distribución y transformación en el medio ambiente

El nitrobenceno se puede degradar por fotolisis y por biodegradación microbiana.

Las propiedades físicas del nitrobenceno parecen indicar que la transferencia del agua al aire será significativa, aunque no rápida. La fotodegradación del nitrobenceno en el aire y el agua es lenta. A partir de los experimentos de fotolisis directa en el aire se determinaron semividas de <1 día, mientras que las semividas calculadas para la reacción con radicales hidroxilo oscilaban entre 19 y 223 días. Con el ozono, la reacción se produce incluso con mayor lentitud. Los experimentos en una cámara de niebla con una mezcla de propileno/butano/dióxido de nitrógeno dieron una semivida estimada para el nitrobenceno de cuatro a cinco días. En masas de agua, la fotolisis directa parece ser la vía de degradación más rápida (semividas de entre 2,5 y 6 días), mientras que la fotolisis indirecta (fotooxidación con radicales hidroxilo, átomos de hidrógeno o electrones hidratados, sensibilización con ácidos húmicos) desempeña una función secundaria (semividas calculadas de 125 días a 13 años para la reacción con los radicales hidroxilo, en función de la concentración del sensibilizador).

Debido a su solubilidad moderada en agua y a la presión de vapor relativamente baja, cabe suponer que la lluvia arrastre en cierta medida el nitrobenceno de la atmósfera; sin embargo, en experimentos sobre el terreno parecía que el arrastre por la lluvia (ya fuera mediante la disolución en las gotas de lluvia o por eliminación del nitrobenceno adsorbido sobre partículas) y unido a partículas era insignificante. Debido a su densidad de vapor (4,1-4,25 veces la del aire), los procesos de eliminación de la atmósfera pueden incluir la sedimentación de vapores.

Los datos reales sobre la evaporación del nitrobenceno a partir de masas de agua parecen ser algo contradictorios, con un pronóstico de las semividas de volatilización mediante un modelo informático de 12 días (ríos) a 68 días (lago eutrófico). La estimación más breve citada en la bibliografía fue de un día (a partir de agua de río); en otro estudio de microcosmos experimentales en los que se simulaba la aplicación de aguas residuales en el suelo se informó de que el nitrobenceno no se había volatilizado, pero se había degradado completamente.

Los estudios de degradación parecen indicar que el nitrobenceno se degrada en las instalaciones de tratamiento de aguas residuales mediante procesos aerobios, con una degradación más lenta en condiciones anaerobias. El nitrobenceno puede no degradarse necesariamente por completo si su concentración en las aguas residuales es alta. Las concentraciones elevadas también pueden inhibir la biodegradación de otros desechos. La biodegradación del nitrobenceno depende sobre todo de la aclimatación de la población microbiana. La degradación mediante inóculos no aclimatados suele ser de muy lenta a insignificante y se produce solamente después de períodos de aclimatación prolongados. Sin embargo, los microorganismos aclimatados, en particular los procedentes de instalaciones industriales de tratamiento de aguas residuales, mostraron una eliminación completa del nitrobenceno en unos días. Se observó que en general la degradación aumentaba en presencia de otros sustratos fácilmente degradables. La adaptación de la microflora y otros sustratos también parecen ser factores limitantes de la descomposición del nitrobenceno en el suelo. Se ha demostrado que la degradación del nitrobenceno en condiciones anaerobias es muy lenta, incluso después de períodos de aclimatación prolongados.

Los factores de bioconcentración medidos para el nitrobenceno en algunos organismos indican un potencial mínimo de bioacumulación; no sufre bioamplificación a través de la cadena alimentaria. Las plantas pueden absorber el nitrobenceno; sin embargo, en los estudios disponibles parecía estar asociado con las raíces y muy poco con otras partes de la planta. En un ecosistema acuático simulado de "estanque piscícola," el nitrobenceno se mantenía fundamentalmente en el agua y no se almacenaba ni amplificaba ecológicamente en la pulga de agua, ni en las larvas de mosquito, los caracoles, las algas o diversos tipos de plancton o de peces.

4.   Niveles ambientales y exposición humana

Las concentraciones de nitrobenceno en muestras del medio ambiente, por ejemplo aguas superficiales, agua freática y aire, son generalmente bajas.

Algunos niveles medidos en el aire de ciudades de los Estados Unidos a comienzos de los años ochenta oscilaban entre <0,05 y 2,1 µg/m³ (<0,01 y 0,41 ppmm) (medias aritméticas). Los datos notificados por la Agencia para la Protección del Medio Ambiente de los Estados Unidos en 1985 indicaban que eran positivas menos del 25% de las muestras de aire en los Estados Unidos, con una concentración mediana de alrededor de 0,05 µg/m³ (0,01 ppmm); en zonas urbanas, los niveles medios eran generalmente inferiores a 1 µg/m³ (0,2 ppmm), con niveles ligeramente más altos en zonas industriales (media de 2,0 µg/m³ [0,40 ppmm]). De 49 muestras de aire medidas en el Japón en 1991, 42 tenían un nivel detectable, de 0,0022-0,16 µg/m³. Los niveles en zonas urbanas y lugares de eliminación de desechos en invierno eran significativamente más bajos (o indetectables) que en verano.

Los datos sobre los niveles de nitrobenceno en el agua superficial parecen ser más amplios que los relativos al aire. Si bien los niveles son variables en función del lugar y la estación, en general se han medido niveles bajos (alrededor de 0,1-1 µg/l). Uno de los niveles más altos notificados fue de 67 µg/l en el río Danubio (Yugoslavia) en 1990. Sin embargo, no se detectó nitrobenceno en ninguna de las muestras de agua superficial recogidas cerca de un gran número de vertederos peligrosos en los Estados Unidos (notificación de 1988). Basándose en los limitados datos, parece que el potencial de contaminación del agua freática podría ser mucho mayor que para el agua superficial; en las mediciones en varios lugares de los Estados Unidos a finales de los años ochenta se detectaron niveles de 210-250 y 1400 µg/l (con niveles mucho más altos en un lugar de gasificación de carbón). En estudios realizados en los años setenta y ochenta en los Estados Unidos y el Reino Unido se ha informado de la presencia de nitrobenceno en el agua de bebida, aunque sólo en una pequeña proporción de muestras, pero no se detectó en 30 muestras canadienses (informe de 1982).

No se localizaron datos sobre la presencia de nitrobenceno en los alimentos, aunque en estudios japoneses realizados en 1991 se detectó en una pequeña proporción de muestras (4 de 147) de peces. No se detectó en un gran número de muestras de biota en un estudio de los Estados Unidos de 1985.

La población general puede estar expuesta a concentraciones variables de nitrobenceno en el aire y posiblemente en el agua de bebida. Hay también una exposición potencial a partir de productos de consumo, pero no se dispone de información precisa. En estudios realizados en el estado de Nueva Jersey, en la costa oriental de los Estados Unidos (con veranos de cálidos a calurosos e inviernos de fríos a muy fríos), las zonas urbanas tenían niveles más altos en verano que en invierno debido a la formación de nitrobenceno por nitración del benceno (procedente del petróleo) y a la mayor volatilidad del nitrobenceno durante los meses más cálidos; la exposición en el aire ambiente en el invierno puede ser insignificante. Basándose en los estudios del aire y en las estimaciones de las emisiones durante la aplicación, son las poblaciones en las cercanías de actividades de fabricación (es decir, productores y consumidores industriales de nitrobenceno para síntesis posterior) y refinerías de petróleo las que probablemente tendrán una exposición significativa al nitrobenceno. Sin embargo, las personas que viven en zonas con residuos peligrosos abandonados y en torno a ellas pueden tener también un potencial de exposición más elevado, debido a la posible contaminación del agua freática y el suelo y a la absorción del nitrobenceno por las plantas.

Los niveles de exposición ocupacional deberían ser inferiores al límite de exposición en el aire ampliamente adoptado de 5 mg/m³ (1 ppm). Basándose en los datos disponibles, parece que el nitrobenceno se absorbe bien por vía cutánea, tanto en forma de vapor como de líquido; por consiguiente, la exposición cutánea puede ser importante, pero no se dispone de datos.

5. Cinética y metabolismo

El nitrobenceno es un líquido volátil que puede penetrar fácilmente en el organismo por inhalación y a través de la piel en forma de vapor, así como por ingestión y absorción cutánea del líquido. La activación del nitrobenceno en ratas para producir metabolitos formadores de metahemoglobina parece estar mediada en un grado importante por la microflora intestinal. En animales de experimentación, la mayor parte del nitrobenceno (alrededor del 80% de la dosis) se metaboliza y elimina en un plazo de tres días. El resto se elimina sólo lentamente. La lentitud del compartimento se debe probablemente a que los eritrocitos reciclan las formas redox del nitrobenceno y los conjugados del glutatión. Se ha demostrado la formación de enlaces covalentes, posiblemente con los grupos sulfhidrilo de la hemoglobina.

Los principales metabolitos urinarios en roedores y conejos son el *p*-nitrofenol y el *p*-amínofenol. En las personas, parte de la dosis absorbida se excreta en la orina; el 10-20% de la dosis se excreta como *p*-nitrofenol (que de esta manera se puede utilizar para la vigilancia biológica). Las semividas de eliminación para el *p*-nitrofenol se estiman en unas cinco horas (fase inicial) y >20 horas (fase posterior). El metabolito urinario *p*-aminofenol es importante sólo con dosis altas.

6. Efectos en mamíferos de laboratorio y en sistemas de prueba *in vitro*

El nitrobenceno produce toxicidad en múltiples órganos por todas las vías de exposición. La exposición oral, cutánea, subcutánea y respiratoria al nitrobenceno en ratones y ratas da lugar a metahemoglobinemia, con la consiguiente anemia hemolítica, congestión esplénica y hematopoyesis en el hígado, la médula ósea y el bazo.

Se observaron en ratas lesiones capsulares esplénicas tanto en la administración mediante sonda (con dosis de sólo 18,75 μg/kg de peso corporal al día) como por vía cutánea (con 100 mg/kg de peso corporal al día o más). Se han observado anteriormente lesiones esplénicas semejantes con colorantes a base de anilina, algunos de los cuales produjeron sarcomas esplénicos en estudios de carcinogenicidad crónica en ratas. Se observaron efectos hepáticos en ratones y ratas tras la administración de nitrobenceno mediante sonda y por vía cutánea, con necrosis de los hepatocitos centrilobulares, aumento de tamaño de los nucleolos hepatocelulares, degeneración hidrópica grave y acumulación de pigmentos en las células de Kupffer. Se detectó un aumento de la vacuolación en la zona X de las glándulas suprarrenales en ratones hembra tras la administración oral y cutánea.

En estudios subcrónicos por vía oral y cutánea en ratones y ratas, se produjeron lesiones del sistema nervioso central en el cerebelo y el tronco encefálico potencialmente mortales. Estas lesiones, con inclusión de hemorragias petequiales, pueden tener efectos tóxicos directos o mediados por efectos vasculares de hipoxia o toxicidad hepática. En función de la dosis, estos efectos neurotóxicos fueron muy manifiestos en forma de ataxia, inclinación de la cabeza y arqueo, pérdida del reflejo de enderezamiento, temblores, coma y convulsiones.

Otros órganos destinatarios eran el riñón (aumento del peso, inflamación del epitelio glomerular y tubular, pigmentación de las células del epitelio tubular), el epitelio nasal (glándularización del epitelio respiratorio, deposición de pigmentos en el epitelio olfatorio y su degeneración), el tiroides (hiperplasia de las células foliculares), el timo (involución) y el páncreas (infiltración de las células mononucleares), mientras que en los conejos se informó de patología pulmonar (enfisema, atelectasis y bronquiolización de las membranas celulares alveolares).

Tras una exposición prolongada por inhalación (505 días) de ratones B6C3F$_1$ macho y hembra, ratas Fischer-344 macho y hembra y ratas Sprague-Dawley macho y hembra, se evaluaron la carcinogenicidad y la toxicidad potenciales del nitrobenceno. La supervivencia no se vio afectada negativamente con las concentraciones sometidas a prueba (hasta 260 mg/m^3 [50 ppm] para los ratones; y hasta 130 mg/m^3 [25 ppm] para las ratas), pero el nitrobenceno inhalado fue tóxico y carcinogénico en ambas especies y en ambas cepas de rata, induciendo en ellas un espectro de neoplasias benignas y malignas (pulmón, tiroides, glándula mamaria, hígado, riñón).

El nitrobenceno no fue genotóxico en bacterias y en células de mamífero *in vitro* y en células de mamífero *in vivo*. Los estudios notificados incluían lesiones en el ADN y valoraciones de reparación, de mutación genética, de efectos cromosomales y de transformación celular.

Numerosos estudios han confirmado la toxicidad testicular del nitrobenceno, siendo los efectos finales espermáticos más sensibles el número y la movilidad de los espermatozoides, seguidos de la movilidad progresiva, la viabilidad, la presencia de esperma anormal y, por último, el índice de fecundidad.

En un estudio de toxicidad reproductiva en dos generaciones con ratas Sprague-Dawley por inhalación, el nitrobenceno a 200 mg/m^3 (40 ppm), pero no a 5 ó 51 mg/m^3 (1 ó 10 ppm), provocó una fuerte disminución del índice de fecundidad de la generaciones F_0 y F_1, asociada con toxicidad del sistema reproductivo masculino; esta disminución de la fecundidad era parcialmente reversible cuando la generación F_1 del grupo tratado con 200 mg/m^3 se emparejaba después de un período de recuperación de nueve semanas con hembras vírgenes no tratadas. Sin embargo, en un estudio de administración oral en la misma raza de ratas (20-100 mg/kg de peso corporal desde 14 días antes del acoplamiento hasta el cuarto día de lactación), el nitrobenceno no tuvo efectos en los parámetros reproductivos, aunque disminuía el peso corporal de las crías y aumentaba la pérdida de peso postnatal. La falta de efectos en la fecundidad en este estudio se debió al reducido intervalo de administración antes del apareamiento y al hecho de que las ratas producen esperma en gran abundancia. Se observó una alteración de la fecundidad masculina con una atrofia testicular importante en ratones y ratas; los efectos en los ratones fueron manifiestos con dosis de 300 mg/kg de peso corporal al día mediante sonda y de 800 mg/kg de peso corporal al día por vía cutánea y en ratas con dosis de 75 mg/kg de peso corporal al día mediante sonda y de 400 mg/kg de peso corporal al día por vía cutánea. Se detectó toxicidad testicular en forma de descamación del epitelio seminífero, aparición de células gigantes multinucleadas, atrofia elevada y aspermia prolongada. El nitrobenceno tiene efectos directos en los testículos, puestos de manifiesto en estudios *in vivo* e *in vitro*. Afecta a la espermatogénesis, con la exfoliación de las células germinales predominantemente viables y la degeneración de las células de Sertoli. El principal efectos histopatológico es la degeneración de los espermatocitos.

En general, no se vieron afectados los órganos reproductivos maternos, excepto en un estudio en el que se observó atrofia uterina en ratones tras una dosis cutánea de 800 mg/kg de peso corporal al día.

Los estudios de toxicidad en el desarrollo en ratas y ratones pusieron de manifiesto que la exposición por inhalación a concentraciones de nitrobenceno suficientes para producir toxicidad materna no provocaba efectos fetotóxicos, embriotóxicos o teratogénicos. Con la concentración más alta sometida a prueba en estos estudios (530 mg/m^3 [104 ppm] en un estudio con ratones), los valores medios de los lugares de resorción y el porcentaje de resorciones/implantaciones fueron más altos en este grupo que en los testigos correspondientes, pero se mantuvieron dentro de la gama histórica de los testigos; se observaron efectos maternos (es decir, mayores niveles de metahemoglobina y aumento del peso del hígado) a partir de 210 mg/m^3 (41 ppm).

En un estudio sobre la inmunotoxicidad del nitrobenceno en ratones B6C3F$_1$, provocó un aumento de la celularidad del bazo, cierto grado de inmunosupresión (disminución de la respuesta de la IgM a los eritrocitos de oveja) y estimulación de la médula ósea. El nitrobenceno no influyó de manera significativa en la resistencia del huésped a la infección microbiana o vírica, aunque había una tendencia a una mayor susceptibilidad cuando la función de las células T contribuía a la defensa del huésped.

7.   Efectos en el ser humano

El nitrobenceno es tóxico para las personas mediante exposición respiratoria, cutánea y oral. El principal efecto sistémico asociado con la exposición humana al nitrobenceno es la metahemoglobinemia.

Se han notificado numerosos casos de intoxicación y muerte accidentales de personas por ingestión de nitrobenceno. En los casos de ingestión oral o cuando los pacientes estaban aparentemente cerca de la muerte debido a una metahemoglobinemia grave, la interrupción de la exposición y una intervención médica pronta determinaban una mejoría gradual y la recuperación. Aunque la exposición humana a cantidades suficientemente altas de nitrobenceno puede ser letal por cualquiera de las vías de exposición, se considera poco probable que se pueda producir la exposición a niveles tan altos que puedan provocar la muerte, excepto en casos de accidentes industriales o suicidios.

Durante la exposición humana al nitrobenceno probablemente es el bazo el órgano destinatario; en una mujer profesionalmente expuesta al nitrobenceno de pintura (principalmente por inhalación), el bazo se reblandeció y aumentó de tamaño.

Se ha informado de efectos en el hígado, en particular aumento de tamaño y reblandecimiento y alteración de la química del suero, en una mujer expuesta al nitrobenceno por inhalación.

Entre los síntomas neurotóxicos notificados en personas tras la exposición por inhalación al nitrobenceno caben mencionar dolor cabeza, confusión, vértigo y náuseas. Entre los efectos en personas expuestas por vía oral figuraban también esos síntomas, así como apnea y coma.

8. Efectos en los organismos en el medio ambiente

El nitrobenceno parece ser tóxico para las bacterias y puede afectar negativamente a las instalaciones de tratamiento de aguas residuales si está presente en concentraciones elevadas en las de entrada. La concentración tóxica más baja notificada para microorganismos es para la bacteria *Nitrosomonas*, con una CE_{50} de 0,92 mg/l basada en la inhibición del consumo de amoníaco. Otros valores notificados son una concentración sin efectos observados (NOEC) a las 72 horas de 1,9 mg/l para el protozoo *Entosiphon sulcatum* y una concentración más baja con efectos observados (LOEC) a los ocho días de 1,9 mg/l para el alga cianofícea *Microcystis aeruginosa*.

En los invertebrados de agua dulce, la toxicidad aguda (valores de la CL_{50} a las 24-48 horas) osciló entre 24 mg/l para la pulga de agua (*Daphnia magna*) y 140 mg/l para el caracol (*Lymnaea stagnalis*). En los invertebrados marinos, el valor más bajo de la toxicidad aguda fue una CL_{50} a las 96 horas de 6,7 mg/l para el mísido *Mysidopsis bahia*. El valor más bajo notificado para las pruebas crónicas fue una NOEC a los 20 días de 1,9 mg/l para *Daphnia magna*, con una CE_{50} basada en la reproducción de 10 mg/l.

Los peces de agua dulce mostraron una sensibilidad igualmente baja al nitrobenceno. Los valores de la CL_{50} a las 96 horas oscilaron entre 24 mg/l para el medaka (*Oryzias latipes*) y 142 mg/l para el guppy (*Poecilia reticulata*). No se observaron efectos en la mortalidad o el comportamiento de medaka con 7,6 mg/l durante una exposición 18 días.

9. Evaluación del peligro y del riesgo

En las personas expuestas se ha observado metahemoglobinemia y posteriormente cambios hematológicos y esplénicos, pero los datos no permiten cuantificar la relación exposición-respuesta. En roedores se observaron metahemoglobinemia, efectos hematológicos, efectos testiculares y, en los estudios de inhalación, efectos en el sistema respiratorio con las dosis más bajas sometidas a prueba. Con el nivel de exposición más bajo estudiado, 5 mg/m^3 (1 ppm), se detectaron en ratas metahemoglobinemia, hipospermia epididimal bilateral y atrofia testicular bilateral. En ratones hubo un aumento relacionado con la dosis en la incidencia de bronquialización de las paredes alveolares e hiperplasia alveolar/bronquial con la dosis más baja sometida a prueba, de 26 mg/m^3 (5 ppm). Se observó una respuesta carcinogénica tras la exposición de ratas y ratones al nitrobenceno: se detectaron adeno-carcinomas mamarios en ratones $B6C3F_1$ hembra, carcinomas hepáticos en ratas Fischer 344 macho y adenocarcinomas de las células foliculares del tiroides en ratas Fischer 344 macho. Se observaron tumores benignos en cinco órganos. Los estudios sobre la geno-toxicidad han dado generalmente resultados negativos.

Aunque hay varios productos metabólicos del nitrobenceno que podrían intervenir en la causalidad del cáncer, no se conoce el mecanismo de acción carcinogénica. Debido a que probablemente los mecanismos redox tienen características comunes en los animales de experimentación y en las personas, se supone que el nitrobenceno puede provocar cáncer en las personas por cualquiera de las vías de exposición.

La exposición de la población general al nitrobenceno a partir del aire o el agua de bebida probablemente es muy baja. Aunque no se pudo obtener ningún valor para la concentración sin efectos adversos observados (NOAEL) a partir de ninguno de los estudios toxicológicos, hay un riesgo al parecer escaso para los efectos no neoplásicos. Si los valores de la exposición son suficientemente bajos para evitar los efectos no neoplásicos, cabe esperar que no se produzcan efectos carcinogénicos.

En el pasado se han producido con frecuencia intoxicaciones agudas por la presencia de nitrobenceno en los productos de consumo. Es posible una exposición humana significativa, debido a la presión de vapor moderada del nitrobenceno y a la importante absorción cutánea. Además, su olor relativamente agradable a almendra puede no desalentar a las personas de consumir alimentos o agua contaminados. Los niños pequeños son especialmente susceptibles a los efectos del nitrobenceno.

Es limitada la información sobre la exposición en el lugar de trabajo. En un estudio en lugares de trabajo, las concentraciones de exposición fueron del mismo orden de magnitud que las concentraciones más bajas con efectos adversos observados (LOAEL) obtenidas en un estudio de inhalación prolongado. Por consiguiente, hay una preocupación importante por la salud de los trabajadores expuestos al nitrobenceno.

El nitrobenceno muestra una escasa tendencia a la bioacumulación y parece experimentar biotransformación tanto aerobia como anaerobia. Con respecto a los sistemas terrestres es poco probable que se den en la naturaleza los niveles de preocupación notificados en las pruebas de laboratorio, excepto posiblemente en zonas próximas a lugares de producción y uso de nitrobenceno y en zonas contaminadas por derrames.

Utilizando los datos disponibles de la toxicidad aguda y un método de distribución estadística, junto con una razón de toxicidad aguda:crónica obtenida a partir de los datos sobre crustáceos, el límite de concentración para el nitrobenceno, a fin de proteger el 95% de las especies de agua dulce con una confianza del 50%, se puede estimar en 200 µg/l. Así pues, es poco probable que, con los niveles normalmente notificados en las aguas superficiales, de alrededor de

0,1-1 µg/l, el nitrobenceno represente un peligro ambiental para las especies acuáticas. Incluso con las concentraciones más altas notificadas (67 µg/l), es poco probable que el nitrobenceno sea motivo de preocupación para las especies de agua dulce.

No se dispone de información suficiente que permita obtener un valor indicativo para los organismos marinos.

THE ENVIRONMENTAL HEALTH CRITERIA SERIES (continued)

Ethylene oxide (No. 55, 1985)
Extremely low frequency (ELF) fields
(No. 36, 1984)
Fenitrothion (No. 133, 1992)
Fenvalerate (No. 95, 1990)
Flame retardants: a general introduction
(No. 192, 1997)
Flame retardants: tris(chloropropyl)
phosphate and tris(2-chloroethyl)
phosphate (No. 209, 1998)
Flame retardants: tris(2-butoxyethyl)
phosphate, tris(2-ethylhexyl) phosphate
and tetrakis(hydroxymethyl)
phosphonium
salts (No. 218, 2000)
Fluorides (No. 227, 2001)
Fluorine and fluorides (No. 36, 1984)
Food additives and contaminants in food,
principles for the safety assessment of
(No. 70, 1987)
Formaldehyde (No. 89, 1989)
Fumonisin B_1 (No. 219, 2000)
Genetic effects in human populations,
guidelines for the study of (No. 46, 1985)
Glyphosate (No. 159, 1994)
Guidance values for human
exposure limits (No. 170, 1994)
Heptachlor (No. 38, 1984)
Hexachlorobenzene (No. 195, 1997)
Hexachlorobutadiene (No. 156, 1994)
Alpha- and beta-hexachlorocyclohexanes
(No. 123, 1992)
Hexachlorocyclopentadiene
(No. 120, 1991)
n-Hexane (No. 122, 1991)
Human exposure assessment
(No. 214, 2000)
Hydrazine (No. 68, 1987)
Hydrogen sulfide (No. 19, 1981)
Hydroquinone (No. 157, 1994)
Immunotoxicity associated with exposure
to chemicals, principles and methods for
assessment (No. 180, 1996)
Infancy and early childhood, principles for
evaluating health risks from chemicals
during (No. 59, 1986)
Isobenzan (No. 129, 1991)
Isophorone (No. 174, 1995)
Kelevan (No. 66, 1986)
Lasers and optical radiation (No. 23,
1982)
Lead (No. 3, 1977)[a]
Lead, inorganic (No. 165, 1995)
Lead – environmental aspects
(No. 85, 1989)
Lindane (No. 124, 1991)
Linear alkylbenzene sulfonates
and related
compounds (No. 169, 1996)
Magnetic fields (No. 69, 1987)
Man-made mineral fibres (No. 77, 1988)
Manganese (No. 17, 1981)
Mercury (No. 1, 1976)[a]
Mercury – environmental aspects
(No. 86, 1989)
Mercury, inorganic (No. 118, 1991)
Methanol (No. 196, 1997)

Methomyl (No. 178, 1996)
2-Methoxyethanol, 2-ethoxyethanol, and
their acetates (No. 115, 1990)
Methyl bromide (No. 166, 1995)
Methylene chloride
(No. 32, 1984, 1st edition)
(No. 164, 1996, 2nd edition)
Methyl ethyl ketone (No. 143, 1992)
Methyl isobutyl ketone (No. 117, 1990)
Methylmercury (No. 101, 1990)
Methyl parathion (No. 145, 1992)
Methyl tertiary-butyl ether (No. 206, 1998)
Mirex (No. 44, 1984)
Morpholine (No. 179, 1996)
Mutagenic and carcinogenic chemicals,
guide to short-term tests for detecting
(No. 51, 1985)
Mycotoxins (No. 11, 1979)
Mycotoxins, selected: ochratoxins,
trichothecenes, ergot (No. 105, 1990)
Nephrotoxicity associated with exposure
to chemicals, principles and methods for
the assessment of (No. 119, 1991)
Neurotoxicity associated with exposure to
chemicals, principles and methods for the
assessment of (No. 60, 1986)
Neurotoxicity risk assessment for human
health, principles and approaches
(No. 223, 2001)
Nickel (No. 108, 1991)
Nitrates, nitrites, and N-nitroso
compounds
(No. 5, 1978)[a]
Nitrogen oxides
(No. 4, 1977, 1st edition)[a]
(No. 188, 1997, 2nd edition)
2-Nitropropane (No. 138, 1992)
Nitro-and nitro-oxypolycyclic aromatic
hydrocarbons, selected (No.229, 2003)
Noise (No. 12, 1980)[a]
Organophosphorus insecticides:
a general introduction (No. 63, 1986)
Palladium (No. 226, 2001)
Paraquat and diquat (No. 39, 1984)
Pentachlorophenol (No. 71, 1987)
Permethrin (No. 94, 1990)
Pesticide residues in food, principles for
the toxicological assessment of
(No. 104, 1990)
Petroleum products, selected
(No. 20, 1982)
Phenol (No. 161, 1994)
d-Phenothrin (No. 96, 1990)
Phosgene (No. 193, 1997)
Phosphine and selected metal
phosphides
(No. 73, 1988)
Photochemical oxidants (No. 7, 1978)
Platinum (No. 125, 1991)
Polybrominated biphenyls (No. 152, 1994)
Polybrominated dibenzo-p-dioxins and
dibenzofurans (No. 205, 1998)
Polychlorinated biphenyls and terphenyls
(No. 2, 1976, 1st edition)[a]
(No. 140, 1992, 2nd edition)
Polychlorinated dibenzo-p-dioxins and
dibenzofurans (No. 88, 1989)

Continued at end of book

THE ENVIRONMENTAL HEALTH CRITERIA SERIES (continued)

Polycyclic aromatic hydrocarbons, selected non-heterocyclic (No. 202, 1998)
Progeny, principles for evaluating health risks associated with exposure to chemicals during pregnancy (No. 30, 1984)
1-Propanol (No. 102, 1990)
2-Propanol (No. 103, 1990)
Propachlor (No. 147, 1993)
Propylene oxide (No. 56, 1985)
Pyrrolizidine alkaloids (No. 80, 1988)
Quintozene (No. 41, 1984)
Quality management for chemical safety testing (No. 141, 1992)
Radiofrequency and microwaves (No. 16, 1981)
Radionuclides, selected (No. 25, 1983)
Reproduction, principles for evaluating health risks associated with exposure to chemicals (No. 225, 2001)
Resmethrins (No. 92, 1989)
Synthetic organic fibres, selected (No. 151, 1993)
Selenium (No. 58, 1986)
Styrene (No. 26, 1983)
Sulfur oxides and suspended particulate matter (No. 8, 1979)
Tecnazene (No. 42, 1984)
Tetrabromobisphenol A and derivatives (No. 172, 1995)
Tetrachloroethylene (No. 31, 1984)
Tetradifon (No. 67, 1986)
Tetramethrin (No. 98, 1990)

Thallium (No. 182, 1996)
Thiocarbamate pesticides: a general introduction (No. 76, 1988)
Tin and organotin compounds (No. 15, 1980)
Titanium (No. 24, 1982)
Tobacco use and exposure to other agents (No. 211, 1999)
Toluene (No. 52, 1986)
Toluene diisocyanates (No. 75, 1987)
Toxicity of chemicals (Part 1), principles and methods for evaluating the (No. 6, 1978)
Toxicokinetic studies, principles of (No. 57, 1986)
Tributyl phosphate (No. 112, 1991)
Tributyltin compounds (No. 116, 1990)
Trichlorfon (No. 132, 1992)
1,1,1-Trichloroethane (No. 136, 1992)
Trichloroethylene (No. 50, 1985)
Tricresyl phosphate (No. 110, 1990)
Triphenyl phosphate (No. 111, 1991)
Tris- and bis(2,3-dibromopropyl) phosphate (No. 173, 1995)
Ultrasound (No. 22, 1982)
Ultraviolet radiation (No. 14, 1979, 1st edition) (No. 160, 1994, 2nd edition)
Vanadium (No. 81, 1988)
Vinyl chloride (No. 215, 1999)
Vinylidene chloride (No. 100, 1990)
White spirit (No. 187, 1996)
Xylenes (No. 190, 1997)
Zinc (No. 221, 2001)

THE CONCISE INTERNATIONAL CHEMICAL ASSESSMENT SERIES

CICADs are IPCS risk assessment documents that provide concise but critical summaries of the relevant scientific information concerning the potential effects of chemicals upon human health and/or the environment

Acrolein (No. 43, 2002)
Acrylonitrile (No. 39, 2002)
Azodicarbonamide (No. 16, 1999)
Arsine: human health aspects (No. 47, 2002)
Barium and barium compounds (No.33, 2001)
Benzoic acid and sodium benzoate (No. 26, 2000)
Benzyl butyl phthalate (No. 17, 1999)
Beryllium and beryllium compounds (No. 32, 2001)
Biphenyl (No. 6, 1999)
Bromoethane (No. 42, 2002)
1,3-Butadiene (No. 30, 2001)
2-Butoxyethanol (No. 10, 1998)
Carbon disulfide (No. 46, 2002)
Chloral hydrate (No. 25, 2000)
Chlorinated naphthalenes (No. 34, 2001)

Chlorine dioxide (No. 37, 2002)
4-Chloroanaline (No.48, 2003)
Crystalline silica, quartz (No. 24, 2000)
Cumene (No. 18, 1999)
1,2-Diaminoethane (No. 15, 1999)
3,3'-Dichlorobenzidine (No. 2, 1998)
1,2-Dichloroethane (No. 1, 1998)
2,2-Dichloro-1,1,1-trifluoroethane
1,1-Dichloroethene (Vinylidene chloride) (no. 51, 2003)
Diethyl phthalate (No. 52, 2003)
Diethylene glycol dimethyl ether (No.41, 2002)
Dimethylformamide (No. 31, 2001)
Diphenylmethane diisocyanate (MDI) (No. 27, 2001)
Ethylenediamine (No. 15, 1999)
Ethylene glycol: environmental aspects (No. 22, 2000)

The Concise International Chemical Assessment Series (continued)

Ethylene glycol: human health aspects
(No. 45, 2002))

Ethylene oxide (No.54, 2003)

Formaldehyde (No. 40, 2002)

2-Furaldehyde (No. 21, 2000)

HCFC-123 (No. 23, 2000)

Hydrogen sulfide: human health aspects
(No. 53, 2003)

Limonene (No. 5, 1998)

Manganese and its compounds
(No. 12, 1999)

Mercury, elemental, and inorganic
mercury compounds: human health
aspects (No. 50, 2003)

Methyl and ethyl cyanoacrylates (No. 36, 2001)

Methyl chloride (No. 28, 2001)

Methyl methacrylate (No. 4, 1998)

Mononitrophenols (No. 20, 2000)

N-Nitrosodimethylamine (No. 38, 2002)

Phenylhydrazine (No. 19, 2000)

N-Phenyl-1-naphthylamine (No. 9, 1998)

Polychlorinated Biphenyls: human health
aspects (No. 55, 2003)

1,1,2,2-Tetrachloroethane (No. 3, 1998)

1,1,2,2-Tetrafluoroethane (No. 11, 1999)

Thiourea (No. 49, 2003)

o-Toluidine (No. 7, 1998)

Tributylin oxide (No. 14, 1999)

1,2,3-Trichloropropane (No.56, 2003)

Triglycidyl isocyanurate (No. 8, 1998)

Triphenyltin compounds (No. 13, 1999)

Vanadium pentoxide and other vanadium
compounds (No. 29, 2001)

To order further copies of monographs in these series, please contact Marketing and Dissemination, World Health Organization, 1211 Geneva 27, Switzerland (Fax: +41-22-791 4857; E-mail: bookorders@who.int)

---

[a] Out of print

www.ingramcontent.com/pod-product-compliance
Lightning Source LLC
Chambersburg PA
CBHW031841200326
41597CB00012B/228